Defining Soil Quality for
a Sustainable Environment

CONTENTS

FOREWORD

Scientists and lay persons have long recognized that the quality of two great natural resources—air and water—can be degraded by human activity. Unfortunately, few people have considered that the quality of soil can also be affected by differing uses and management practices. Interest in *soil quality* has heightened during the past three years as a small cadre of soil scientists became more concerned about the role of soils in sustainable production systems and the linkages between soil characteristics and plant–human health.

The concept of *soil quality* will not be in the mainstream of soil or environmental science programs until there is wide acceptance of the definition for the term and quantitative indicators of *soil quality* are developed. Air and water quality are well recognized concepts that have standards established by law and regulation. A great deal of study and education will be necessary before *soil quality* becomes an important national natural resource issue.

A symposium entitled "Defining Soil Quality for a Sustainable Environment" was held at the 1992 annual meeting of the Tri-societies as an initial attempt at dialogue on soil quality within the Soil Science Society of America. This publication contains the papers on soil quality presented at the symposium. Emphasis has been placed on defining soil quality, identifying soil quality indices, and assessing the biological importance of soil quality. The publication represents an excellent first step in educating scientists about the concepts and significance of soil quality.

DARRELL W. NELSON, *president*
Soil Science Society of America

PREFACE

For humankind, soil is the essence of life and health. As with air and water, human life could not be sustained without the soil because it is the source of most of our food. The economic well-being of the USA and most other nations depends greatly on the soil and how well its productivity is maintained. However, soil does far more for society than just produce food. A healthy or good quality soil acts as an environmental filter for cleansing air and water. It is a major sink for global gases and if managed properly can favorably affect the carbon dioxide balance, which is important for combating global climate change. Soil is the ultimate receptor and incubation chamber for the decomposition and detoxification of organic wastes, and recycling of nutrients from these materials back to plants. If mismanaged, the soil can work against us; it can pollute the air and water and cease producing abundant and nutritious food. Many people believe that, within a community, there is a strong link between the health or quality of the soil, food quantity and quality, and the health, well-being and prosperity of the people who live there.

For many years the USA and other nations have sought policies to protect their agricultural soils against degradation and to improve them to ensure sustainable food production for future generations. Principles of soil conservation have been known for centuries and national conservation programs to safeguard and preserve agricultural lands have been implemented by governments of various countries since the 1930s. Yet recent assessments conducted on regional and global scales indicate that the ravages of human-induced degradation (soil erosion, salinization, organic matter decline, etc.) are causing the loss of millions of hectares of agricultural land every year. Are not the positive interventions by human action having an impact on soil rehabilitation, or are these simply being overwhelmed or masked by the large-scale assessments of degradation processes? In addition to assessments of degradation, a more quantitative assessment is needed of how farming practices are affecting the capacity of the soil to produce food and perform certain environmental functions (i.e., soil quality) and whether the capacity is being degraded, aggraded, or is remaining unchanged. Often the gross benefits to the soil of *improved* management practices such as increased productivity, water retention, and resistance to erosion occur slowly and are not measurable in the early years of treatment. As a result, soil conservation and soil rehabilitation programs may be abandoned prematurely or lack support, because it is not possible, early on, to determine their potential economic worth in terms of soil improvement. Even where management practices have been imposed over longer periods methodology has not been available to directly quantify the degree of change with respect to certain functions the soil is expected to perform.

Take, for example, the Conservation Reserve Program (CRP) and the Conservation Compliance Provisions of the 1985 Food Security Act. The CRP enables farmers to bid out their highly erodible lands and receive payment from the U.S. government in return for maintaining these lands in grass, tree cover, or both for 10 years. Conservation compliance requires that highly erodible lands must be farmed according to an approved conservation plan for the farmer to be eligible for Government Farm Program benefits. Both programs require a sizeable investment from society, which in turn expects payoff, at least in part, from maintenance and improvement of the soil resource base for future use. But how can changes in the soil condition be evaluated so as to determine whether these policies are achieving an improvement in the soil condition, and if so, by how much and what is the economic worth to society? Through economic analyses like natural resource accounting, these improvements in the soil, if measured, can then be factored in or given proper credits.

There is new emphasis on soil quality as a more sensitive and dynamic way to document the condition of our soils, how they respond to management changes, and their resilience to stresses imposed by natural forces or farming practices. It was toward this goal that the Rodale Institute Research Center took the lead in sponsoring and convening a workshop, "Assessment and Monitoring of Soil Quality," 11–13 July 1991, in Emmaus, PA. A concensus of the workshop was that soil quality should not be limited to soil productivity, but should encompass environmental quality, human and animal health, and food safety and quality. Possibilities for developing a soil quality index was explored. It was also recognized that past emphasis was given to physical and chemical properties as indicators of soil quality rather than to biological properties, which are more difficult to measure, predict, or quantify. However, biological and ecological-based indicators of soil quality are believed more dynamic than those based on physical and chemical properties, and therefore, have the advantage of serving as early signals of soil degradation, or of aggradation in the case of soil improvement.

Since the workshop in Emmaus, many new projects on soil quality have been initiated, and some on-going projects have begun to coordinate efforts among scientists, both nationally and internationally. It will require the concerted efforts of scientists from many disciplines, working with farmers, to achieve our goal of developing a realistic, useful, soil quality index.

This publication is comprised of papers from a symposium on soil quality held at the 1992 American Society of Agronomy annual meeting in Minneapolis, MN. The objectives were to identify the major components of soil that define soil quality and to quantify soil-derived indicators of soil quality with major emphasis on soil biology. Themes addressed by the authors include approaches to defining and assessing soil quality relationships between various soil properties and soil quality and land-use sustainability, models for characterizing soil quality, and defining biological criteria for evaluating the effects of management practices on soil quality. Several papers relate to measurement of specific soil quality components.

The papers in these proceedings contribute new perspectives on the definition of soil quality and how it may be measured and quantified. It is hoped that the knowledge presented here will help lead toward national and global soil quality monitoring and assessment programs as a strategy to help preserve and enhance the long-term use of the valuable soil resource base for food and environmental needs. Reliable indicators will make it possible to evaluate the long-term impact of land use and management on soil and environmental quality. only then will farmers, scientists, extension specialists, policy-makers, and the public have the necessary fore hand knowledge to understand measures to protect and preserve the health of our soil resource base for sustainable land-use.

RHONDA R. JANKE
Rodale Institute Research Center
Kutztown, PA

ROBERT I. PAPENDICK
USDA-ARS, Washington State University
Pullman, WA

PROLOGUE

Meaningful change occurs slowly. The first step and lengthiest ordeal usually involves convincing responsible people that change is necessary, desirable, and possible. In social causes, this is known as *raising awareness*. Since its inception, first as the Soil and Health Foundation in 1947 and now in its present form, the Rodale Institute has worked toward raising awareness among agricultural researchers, policy makers, and farmers regarding the importance of soil quality and its relationship to human health issues. We are pleased by the increasing convergence of thought and attention on the important role of soil in the natural environment.

The 18 manuscripts and extended papers published here by some of today's finest soil and agricultural researchers is evidence that meaningful positive change is occurring, that soil quality issues have been recognized—outside of the Rodale circle—as bearing equal environmental impact to water and air quality concerns in the USA and abroad. The first step is safely on its way. Now we must consider how to apply what we have learned so far, and what we will do with each newly discovered *truth*, because knowledge without application is frequently a futile exercise.

It is my fervent hope that these thoughtful papers will help propel those of us grappling with soil quality issues toward the next step, the creation of an international soil quality index. We need to know the state of the world's soils and then to document changes in that status on a regular basis if we are to protect the soil we all depend on for life. I believe that when this is accomplished we will be able to achieve, through public awareness, an appreciation of soil quality and its contribution to human health and well-being.

JOHN HABERERN, *president*
Rodale Institute
Emmaus, PA

CONTRIBUTORS

D. F. Bezdicek — Director, Center for Sustaining Agriculture and Natural Resources, Washington State University, Pullman, WA 99164-6240

Robert L. Blevins — Professor, Department of Agronomy, University of Kentucky, Lexington, KY 40546-0091

Patrick J. Bohlen — Doctoral Candidate, Department of Entomology, Ohio State University, Columbus, OH 43210

Scott D. Bridgham — Research Associate, Natural Resources Research Institute, University of Minnesota, Duluth, MN 55811

Ingrid C. Burke — Assistant Professor, Department of Forest Sciences, Colorado State University, Fort Collins, CO 80523

R. Chaussod — Research Scientist, Laboratoire des Sols, Institut National de la Recherche Agronomique (INRA), 21034 Dijon, France

H. H. Cheng — Professor, Department of Soil Science, University of Minnesota, St. Paul, MN 55108

C. Vernon Cole — Soil Scientist, USDA-ARS, Natural Resource Ecology Laboratory, Colorado State University, Fort Collins, CO 80523

David C. Coleman — Professor of Ecology, Ecology/Biosciences, Institute of Ecology, University of Georgia, Athens, GA 30602-2602

Harold P. Collins — Research Associate, Kellogg Biological Station, Michigan State University, Hickory Corners, MI 49060-9516

Richard P. Dick — Associate Professor of Soil Science, Department of Soil Science, Oregon State University, Corvallis, OR 97331

John W. Doran — Soil Scientist, USDA-ARS, University of Nebraska, Lincoln, NE 68583

John M. Duxbury — Professor of Soil Science, Cornell University, Ithaca, NY 14853

Neal S. Eash — Research Associate, Iowa State University, Ames, IA 50011

Clive A. Edwards — Professor of Environmental Science, Department of Entomology, Ohio State University, Columbus, OH 43210

Edward T. Elliott — Senior Research Scientist and Associate Director, Natural Resource Ecology Laboratory, Colorado State University, Fort Collins, CO 80523

Mary F. Fauci — Research Assistant, Oregon State University. Currently Research Associate, Department of Crop and Soil Science, Washington State University, Pullman, WA 99164-6420

Serita D. Frey — Research Associate, Natural Resource Ecology Laboratory, Colorado State University, Fort Collins, CO 80523

Wilbur W. Frye	Professor of Agronomy and Director of Regulatory Services, University of Kentucky, Lexington, KY 40546-0275
Fernando O. Garcia	Research Scientist, Institut Nacional de Technologia Agropecuaria, Estacion Experimental Agropecuaria, C.C. 276-7620 Balcarce, Argentina
M. J. Garlynd	Research Assistant, Department of Soil Science, University of Wisconsin, Madison, WI 53706-1299
Peter R. Grace	Research Fellow, CRC for Soil and Land Management, Glen Osmond, South Australia 5064
V. V. S. R. Gupta	Research Scientist, Division of Plant Industry, CSIRO, Canberra, ACT 2601, Australia. Currently Research Fellow, CRC for Soil and Land Management, Glen Osmond, South Australia 5064
Ardell D. Halvorson	Soil Scientist-Research Leader, USDA-ARS, Central Great Plains Research Station, Akron, CO 80720
Jonathan J. Halvorson	Research Associate/Ecologist, USDA-ARS/NORCUS, Washington State University, Pullman, WA 99164
R. F. Harris	Professor, Department of Soil Science, University of Wisconsin, Madison, WI 53706-1299
Paul F. Hendrix	Soil Ecologist, Institute of Ecology and Department of Crop and Soil Sciences, University of Georgia, Athens, GA 30602
Michael V. Hickman	Research Agronomist, USDA-ARS, Botany and Plant Pathology, Purdue University, West Lafayette, IN 47907-1155
Sabine Houot	Research Scientist, Laboratoire des Sols, Institut National de la Recherche Agronomique (INRA), 7885 Thiverval-Grignon, France
David R. Huggins	Assistant Professor of Soil Science, Southwest Experiment Station, University of Minnesota, Lamberton, MN 56152
M. D. Jawson	Chief, Subsurface Processes Branch, USEPA, Robert S. Kerr Environmental Laboratory, Ada, OK 74820
Carol A. Johnston	Senior Research Associate, Natural Resources Research Institute, University of Minnesota, Duluth, MN 55811
Douglas L. Karlen	Research Soil Scientist, USDA-ARS, National Soil Tilth Laboratory, Ames, IA 50011
A. C. Kennedy	Research Scientist, USDA-ARS-LMWC, Washington State University, Pullman, WA 99164-6421
A. V. Kurakov	Professor, Department of Soil Science, Moscow State University, Moscow, Russia
W. E. Larson	Professor Emeritus, Soil Science Department, University of Minnesota, St. Paul, MN 55108
Dennis R. Linden	Soil Scientist, USDA-ARS, Soil Science Department, University of Minnesota, St. Paul, MN 55108

Drew J. Lyon — Assistant Professor of Agronomy, Panhandle Research and Extension Center, University of Nebraska, Scotts Bluff, NE 69361-4939

J. A. E. Molina — Professor, Department of Soil Science, University of Minnesota, St. Paul, MN 55108

Christopher A. Monz — Research Associate, Natural Resource Ecology Laboratory, Colorado State University, Fort Collins, CO 80523. Currently Research Manager, National Outdoor Leadership School, Lander, WY 82520-3128

B. Nicolardot — Research Scientist, Laboratoire des Sols, Institut National de la Recherche Agronomique (INRA), 21034 Dijon, France

Sipho V. Nkambule — Graduate Research Assistant, Cornell University. Currently Lecturer, Crop Production Department, University of Swaziland, P.O. Luyengo, Swaziland

Robert I. Papendick — SOil Scientist, USDA-ARS, Washington State University, Pullman, WA 99164-6421

Timothy B. Parkin — Research Microbiologist, USDA-ARS, National Soil Tilth Laboratory, Ames, IA 50011

John Pastor — Senior Research Associate, Natural Resources Research Institute, University of Minnesota, Duluth, MN 55811

Eldor A. Paul — Chair and Professor, Department of Crop and Soil Science, Michigan State University, East Lansing, MI 48824

Keith H. Paustian — Assistant Professor (Research), Kellogg Biological Station, Michigan State University, Hickory Corners, MI 49060-9516. Currently Research Scientist, Natural Resource Ecology Laboratory, Colorado State University, Fort Collins, CO 80523

F. J. Pierce — Associate Professor, Crop and Soil Sciences Department, Michigan State University, East Lansing, MI 48824

Charles W. Rice — Associate Professor, Department of Agronomy, Kansas State University, Manhattan, KS 66506-5501

D. E. Romig — Research Assistant, Department of Soil Science, University of Wisconsin, Madison, WI 53706-1299

M. M. Roper — Senior Research Scientist, CSIRO, Laboratory for Rural Research, Wembley, Western Australia 6014

Jeffrey L. Smith — Soil Biochemist, USDA-ARS, Washington State University, Pullman, WA 99164-6421

Diane E. Stott — Soil Microbiologist, USDA-ARS, National Soil Erosion Research Laboratory, Purdue University, West Lafayette, IN 47907

Ronald F. Turco — Associate Professor, Department of Agronomy, Laboratory for Soil Microbiology, Purdue University, West Lafayette, IN 47907

Karen Updegraff — Assistant Scientist, Natural Resources Research Institute, University of Minnesota, Duluth, MN 55811

Petra C. J. van Vliet — Soil Ecologist, Institute of Ecology, University of Georgia, Athens, GA 30602

Conversion Factors for SI and non-SI Units

Conversion Factors for SI and non-SI Units

To convert Column 1 into Column 2, multiply by	Column 1 SI Unit	Column 2 non-SI Unit	To convert Column 2 into Column 1, multiply by
Length			
0.621	kilometer, km (10^3 m)	mile, mi	1.609
1.094	meter, m	yard, yd	0.914
3.28	meter, m	foot, ft	0.304
1.0	micrometer, μm (10^{-6} m)	micron, μ	1.0
3.94×10^{-2}	millimeter, mm (10^{-3} m)	inch, in	25.4
10	nanometer, nm (10^{-9} m)	Angstrom, Å	0.1
Area			
2.47	hectare, ha	acre	0.405
247	square kilometer, km^2 (10^3 m)^2	acre	4.05×10^{-3}
0.386	square kilometer, km^2 (10^3 m)^2	square mile, mi^2	2.590
2.47×10^{-4}	square meter, m^2	acre	4.05×10^3
10.76	square meter, m^2	square foot, ft^2	9.29×10^{-2}
1.55×10^{-3}	square millimeter, mm^2 (10^{-3} m)^2	square inch, in^2	645
Volume			
9.73×10^{-3}	cubic meter, m^3	acre-inch	102.8
35.3	cubic meter, m^3	cubic foot, ft^3	2.83×10^{-2}
6.10×10^4	cubic meter, m^3	cubic inch, in^3	1.64×10^{-5}
2.84×10^{-2}	liter, L (10^{-3} m^3)	bushel, bu	35.24
1.057	liter, L (10^{-3} m^3)	quart (liquid), qt	0.946
3.53×10^{-2}	liter, L (10^{-3} m^3)	cubic foot, ft^3	28.3
0.265	liter, L (10^{-3} m^3)	gallon	3.78
33.78	liter, L (10^{-3} m^3)	ounce (fluid), oz	2.96×10^{-2}
2.11	liter, L (10^{-3} m^3)	pint (fluid), pt	0.473

To convert Column 2 into Column 1, multiply by	Column 1 SI Unit	Column 2 non-SI Unit	To convert Column 1 into Column 2, multiply by
Mass			
2.20×10^{-3}	gram, g (10^{-3} kg)	pound, lb	454
3.52×10^{-2}	gram, g (10^{-3} kg)	ounce (avdp), oz	28.4
2.205	kilogram, kg	pound, lb	0.454
0.01	kilogram, kg	quintal (metric), q	100
1.10×10^{-3}	kilogram, kg	ton (2000 lb), ton	907
1.102	megagram, Mg (tonne)	ton (U.S.), ton	0.907
1.102	tonne, t	ton (U.S.), ton	0.907
Yield and Rate			
0.893	kilogram per hectare, kg ha^{-1}	pound per acre, lb acre^{-1}	1.12
7.77×10^{-2}	kilogram per cubic meter, kg m^{-3}	pound per bushel, lb bu^{-1}	12.87
1.49×10^{-2}	kilogram per hectare, kg ha^{-1}	bushel per acre, 60 lb	67.19
1.59×10^{-2}	kilogram per hectare, kg ha^{-1}	bushel per acre, 56 lb	62.71
1.86×10^{-2}	kilogram per hectare, kg ha^{-1}	bushel per acre, 48 lb	53.75
0.107	liter per hectare, L ha^{-1}	gallon per acre	9.35
893	tonnes per hectare, t ha^{-1}	pound per acre, lb acre^{-1}	1.12×10^{-3}
893	megagram per hectare, Mg ha^{-1}	pound per acre, lb acre^{-1}	1.12×10^{-3}
0.446	megagram per hectare, Mg ha^{-1}	ton (2000 lb) per acre, ton acre^{-1}	2.24
2.24	meter per second, m s^{-1}	mile per hour	0.447
Specific Surface			
10	square meter per kilogram, m^2 kg^{-1}	square centimeter per gram, cm^2 g^{-1}	0.1
1000	square meter per kilogram, m^2 kg^{-1}	square millimeter per gram, mm^2 g^{-1}	0.001
Pressure			
9.90	megapascal, MPa (10^6 Pa)	atmosphere	0.101
10	megapascal, MPa (10^6 Pa)	bar	0.1
1.00	megagram per cubic meter, Mg m^{-3}	gram per cubic centimeter, g cm^{-3}	1.00
2.09×10^{-2}	pascal, Pa	pound per square foot, lb ft^{-2}	47.9
1.45×10^{-4}	pascal, Pa	pound per square inch, lb in^{-2}	6.90×10^3

(continued on next page)

Conversion Factors for SI and non-SI Units

To convert Column 1 into Column 2, multiply by	Column 1 SI Unit	Column 2 non-SI Unit	To convert Column 2 into Column 1, multiply by
Temperature			
1.00 (K − 273)	Kelvin, K	Celsius, °C	1.00 (°C + 273)
(9/5 °C) + 32	Celsius, °C	Fahrenheit, °F	5/9 (°F − 32)
Energy, Work, Quantity of Heat			
9.52×10^{-4}	joule, J	British thermal unit, Btu	1.05×10^{3}
0.239	joule, J	calorie, cal	4.19
10^{7}	joule, J	erg	10^{-7}
0.735	joule, J	foot-pound	1.36
2.387×10^{-5}	joule per square meter, $J\ m^{-2}$	calorie per square centimeter (langley)	4.19×10^{4}
10^{5}	newton, N	dyne	10^{-5}
1.43×10^{-3}	watt per square meter, $W\ m^{-2}$	calorie per square centimeter minute (irradiance), $cal\ cm^{-2}\ min^{-1}$	698
Transpiration and Photosynthesis			
3.60×10^{-2}	milligram per square meter second, $mg\ m^{-2}\ s^{-1}$	gram per square decimeter hour, $g\ dm^{-2}\ h^{-1}$	27.8
5.56×10^{-3}	milligram (H_2O) per square meter second, $mg\ m^{-2}\ s^{-1}$	micromole (H_2O) per square centimeter second, $\mu mol\ cm^{-2}\ s^{-1}$	180
10^{-4}	milligram per square meter second, $mg\ m^{-2}\ s^{-1}$	milligram per square centimeter second, $mg\ cm^{-2}\ s^{-1}$	10^{4}
35.97	milligram per square meter second, $mg\ m^{-2}\ s^{-1}$	milligram per square decimeter hour, $mg\ dm^{-2}\ h^{-1}$	2.78×10^{-2}
Plane Angle			
57.3	radian, rad	degrees (angle), °	1.75×10^{-2}

Electrical Conductivity, Electricity, and Magnetism

To convert Column 2 into Column 1, multiply by	Column 1 SI Unit	Column 2 non-SI Unit	To convert Column 1 into Column 2, multiply by
10	siemen per meter, S m^{-1}	millimho per centimeter, mmho cm^{-1}	0.1
10^4	tesla, T	gauss, G	10^{-4}

Water Measurement

9.73 × 10^{-3}	cubic meter, m^3	acre-inches, acre-in	102.8
9.81 × 10^{-3}	cubic meter per hour, m^3 h^{-1}	cubic feet per second, ft^3 s^{-1}	101.9
4.40	cubic meter per hour, m^3 h^{-1}	U.S. gallons per minute, gal min^{-1}	0.227
8.11	hectare-meters, ha-m	acre-feet, acre-ft	0.123
97.28	hectare-meters, ha-m	acre-inches, acre-in	1.03 × 10^{-2}
8.1 × 10^{-2}	hectare-centimeters, ha-cm	acre-feet, acre-ft	12.33

Concentrations

1	centimole per kilogram, cmol kg^{-1} (ion exchange capacity)	milliequivalents per 100 grams, meq 100 g^{-1}	1
0.1	gram per kilogram, g kg^{-1}	percent, %	10
1	milligram per kilogram, mg kg^{-1}	parts per million, ppm	1

Radioactivity

2.7 × 10^{-11}	becquerel, Bq	curie, Ci	3.7 × 10^{10}
2.7 × 10^{-2}	becquerel per kilogram, Bq kg^{-1}	picocurie per gram, pCi g^{-1}	37
100	gray, Gy (absorbed dose)	rad, rd	0.01
100	sievert, Sv (equivalent dose)	rem (roentgen equivalent man)	0.01

Plant Nutrient Conversion

	Elemental	*Oxide*	
2.29	P	P_2O_5	0.437
1.20	K	K_2O	0.830
1.39	Ca	CaO	0.715
1.66	Mg	MgO	0.602

Chapters
1–8

Full-length chapters

Chapters 1–8

1 Defining and Assessing Soil Quality

John W. Doran

USDA-ARS
University of Nebraska
Lincoln, Nebraska

Timothy B. Parkin

USDA-ARS
National Soil Tilth Laboratory
Ames, Iowa

Recent interest in evaluating the quality of our soil resources has been stimulated by increasing awareness that *soil* is a critically important component of the earth's biosphere, functioning not only in the production of food and fiber but also in the maintenance of local, regional, and worldwide environmental quality. A recent call for development of a soil health index was stimulated by the perception that human health and welfare is associated with the quality and health of soils (Haberern, 1992). However, an international conference on assessment and monitoring of soil quality identified that defining and assessing soil quality and health is complicated by the need to consider the multiple functions of soil and to integrate the physical, chemical, and biological soil attributes that define soil function (Papendick & Parr, 1992; Rodale Inst., 1991). The alarming paucity of information on biological indicators of soil quality and methods for integrating physical, chemical, and biological soil properties with soil management practices to assess soil quality led to this special publication, *Defining Soil Quality for a Sustainable Environment.*

The purpose of this chapter is to discuss approaches to defining and assessing soil quality and to suggest one possible form for a soil quality index.

IMPORTANCE OF SOIL FUNCTION

We enter the 21st century with greater awareness of our technological capability to influence the global environment and of the impending crisis for sustaining life on earth (Sagan, 1992; Bhagat, 1990). Increasing concern for sustainable global development was reflected by the participation of heads of state and delegates from 178 countries at the United Nations Conference on Environment & Development, held in Rio de Janeiro in June 1992. Our past management of nature to meet the food and fiber needs of ever-increasing populations has taxed the resiliency of natural processes to maintain global balance and the cycling of energy and matter. Since the 1980s, severe degradation of the soil's productive capacity occurred on more than 10% of the earth's vegetated land as a result of soil erosion, atmospheric pollution, cultivation, over-grazing, land clearing, salinization, and desertification (World Resources Inst., 1992; Sanders, 1992). Drinking and surface and subsurface water quality has been jeopardized in many parts of the world by our choice of land management practices and the consequent imbalance of C and N cycling in soil. The present threat of global climate change and O_3 depletion, through elevated levels of atmospheric gases and altered hydrological cycles, mandates a better understanding of the effects of land management on soil processes. Soil management practices such as tillage, cropping patterns, and fertilization influence water quality, and it was recently shown that such management also influences atmospheric quality through changes in the soil's capacity to produce and/or consume important atmospheric gases such as CO_2, N_2O, and CH_4 (CAST, 1992; Mosier et al., 1991).

Soil is a dynamic, living, natural body that plays many key roles in terrestrial ecosystems. The components of soil include inorganic mineral matter (sand, silt, and clay particles), organic matter, water, gases, and living organisms such as earthworms, insects, bacteria, fungi, algae, and nematodes. There is continual interchange of molecules and ions between the solid, liquid, and gaseous phases that are mediated by physical, chemical, and biological processes. The importance of the microbial component of soil is often overlooked, because it is largely invisible to the naked eye. However, essential parts of the global C, N, P, and S, and water cycles are carried out in soil largely through microbial and faunal interactions with soil physical and chemical properties. Soil organic matter is a major terrestrial pool for C, N, P, and S, and the cycling and availability of these elements are constantly being altered by microbial mineralization and immobilization. Inorganic constituents of soil play a major role in retaining cations (through ion exchange) and nonpolar organic compounds and anions (through sorption reactions). Soil also serves as an essential reservoir of water for terrestrial plants and microorganisms and as a purifying medium through which water passes.

The thin layer of soil covering the earth's surface represents the difference between survival and extinction for most terrestrial life. Soil is a vital natural resource that is nonrenewable on a human time scale (Jenny, 1980). The quality of a soil is largely defined by soil function and represents a composite of its physical, chemical, and biological properties that (i) provide a medium for plant growth, (ii) regulate and partition water flow in the environment, and (iii) serves as an environmental buffer in the formation, attenuation, and degradation of environmentally hazardous compounds (Larson & Pierce, 1991). Soil serves as a medium for plant growth by providing physical support, water, essential nutrients, and oxygen for roots. The suitability of soil for sustaining plant growth and biological activity is a function of physical properties (porosity, water-holding capacity, structure, and tilth) and chemical properties (nutrient supplying ability, pH, salt content, etc.). Many of the soil's biological, physical, and chemical properties are a function of soil organic matter content. Soils play a key role in completing the cycling of major elements required by biological systems, decomposing of organic wastes, and detoxifying certain hazardous compounds. The key role played by soils in recycling organic materials into CO_2 and water and the degrading of chemical pollutants is manifest through microbial decomposition, chemical hydrolysis, complexation, and sorption reactions. The ability of a soil to store and transmit water is a major factor regulating water availability to plants and transport of environmental pollutants to surface and groundwater.

Mechanical cultivation and the continuous growing of row crops resulted in soil loss through erosion, decreases in soil organic matter content, and the concomitant release of CO_2 to the atmosphere (Houghton et al., 1983). Intensive crop production also resulted in excessive loss of topsoil through wind and water erosion. Decreased organic matter content and the use of large tillage and harvesting equipment has decreased structure, tilth, water holding capacity, water infiltration, and increased compaction. Development of saline and sodic soils after initiation of cultivation resulted from both inefficient irrigation techniques and natural processes. In certain areas, improper disposal of hazardous, recalcitrant chemical pollutants contaminated soils, so they are unsuitable for crop production or development and pose a threat to environmental quality and animal health. As a result of the above, we conclude that the quality of many soils in the USA has declined significantly since cultivation was initiated.

The Chinese saying, "The soil is the mother of all things," is a simple statement of the importance of soil to life of all living creatures. Soil function is essential to the sustainability of soil to life of all living creatures. As such, the ability to define and assess soil quality is essential to development, performance, and evaluation of sustainable land and soil management systems. Two important uses for soil quality assessment are (i) as a management tool or aid for farmers and (ii) as a measure of sustainability, of what is happening to our soils, and what we have to leave our grandchildren. We have a responsibility of returning to the soil the vitality it shares with us and to ensure that vitality for generations to come; "As we work our land to produce food, will we leave a legacy of gardens or deserts?" (Haberern, 1992). Our approaches to defining and assessing soil quality should be shaped by these end uses.

DEFINING SOIL QUALITY

Much like air or water, the *quality* of soil has a profound effect on the health and productivity of a given ecosystem and the environments related to it. However, unlike air or water for which we have quality standards, soil quality has been difficult to define and quantify. Many people, including the senior author of this chapter only a short time ago, believe that soil quality is an abstract characteristic of soils that can't be defined, because it depends on external factors such as land use and soil management practices, ecosystem and environmental interactions, socioeconomic and political priorities, and so on. Perceptions of what constitutes a *good* soil vary depending on individual priorities with respect to soil function. However, to manage and maintain our soils in an acceptable state for future generations, *soil quality* must be defined, and the definition must be broad enough to encompass the many facets of soil function.

Attempts to define soil quality should begin with a definition of the word *quality*. According to *Webster's Third New International Dictionary* (Grove, 1986), the word *quality* is a noun derived from the Latin word *qualitas* meaning *of what kind*. Usages and definitions in English include: (i) essential character: NATURE or KIND; (ii) a distinctive inherent feature or attribute: PROPERTY or VIRTUE; (iii) a character position or role: CAPACITY; (iv) degree of excellence: GRADE or CALIBER. *Soil quality* encompasses all these usages of the word *quality* in that it includes the nature and properties or attributes of soil as they relate to the capacity of soil to function effectively. Several definitions of soil quality that have recently been proposed are as follows:

Soil qualities—Inherent attributes of soils that are inferred from soil characteristics or indirect observations (e.g., compactibility, erodibility, and fertility).
(SSSA, 1987)

The ability of soil to support crop growth which includes factors such as degree of tilth, aggregation, organic matter content, soil depth, water holding capacity, infiltration rate, pH changes, nutrient capacity, and so forth.
(Power & Myers, 1989)

The capacity of a soil to function in a productive and sustained manner while maintaining or improving the resource base, environment, and plant, animal, and human health.
(NCR-59 meeting minutes, Madison, WI, September, 1991)

The capacity of a soil to function within the ecosystem boundaries and interact positively with the environment external to that ecosystem.
(Larson & Pierce, 1991)

The capability of soil to produce safe and nutritious crops in a sustained manner over the long-term, and to enhance human and animal health, without impairing the natural resource base or harming the environment.
(Parr et al., 1992)

Simply put: "Fitness for use."   (Pierce & Larson, 1993)

Common to all these definitions of soil quality is the capacity of soil to function effectively at present and in the future. Confusion as to what *soil quality* means often results from failure to identify the major issues of concern with respect to soil function. These issues were identified at a recent conference on assessment and monitoring of soil quality (Rodale Inst., 1991) as:

Fig. 1-1. Major issues or components that define soil quality.

1. Productivity—the ability of soil to enhance plant and biological productivity.
2. Environmental quality—the ability of soil to attenuate environmental contaminants, pathogens, and offsite damage.
3. Animal health—the interrelationship between soil quality and plant, animal, and human health (see Fig. 1-1).

We propose in this chapter to define soil quality as follows:

The capacity of a soil to function within ecosystem boundaries to sustain biological productivity, maintain environmental quality, and promote plant and animal health.

ASSESSMENT OF SOIL QUALITY

Our ability to assess soil quality and identify key soil properties that serve as indicators of soil function is complicated by the many issues defining quality and the multiplicity of physical, chemical, and biological factors that control biogeochemical processes and their variation in time, space, and intensity. Practical assessment of soil quality requires consideration of these functions and their variations in time and space (Larson & Pierce, 1991). Soil quality assessment, however, is invaluable to determining the sustainability of land management systems in the near and distant future. A soil quality index is needed to identify problem production areas, make realistic estimates of food production, monitor changes in sustainability and environmental quality as related to agricultural management, and assist federal and state agencies in formulating and evaluating sustainable agricultural and land-use policies (Granatstein & Bezdicek, 1992). Of particular note is *how soil quality assessment can be used to evaluate the benefits from public investment in farm policy programs.*

Land management is sustainable only when it maintains or improves resource quality, specifically the quality of air, soil, water, and food resources. Soil quality assessment provides a basic means to evaluate the sustainability of agricultural and land management systems. Soils have various levels of quality that are basically defined by stable natural or inherent features related to soil-forming factors and dynamic changes induced by soil management (Pierce & Larson, 1993). Detecting changes in the dynamic component of soil quality is essential to evaluating the performance and sustainability of soil management systems. Pierce and Larson (1993) recently proposed an approach based on establishing a minimum data set of temporarily variable soil properties and pedotransfer functions. Pedotransfer functions serve to relate soil characteristics and properties with each other in evaluating soil quality and also in estimating soil attributes that are difficult to measure. This approach relies on available soil surveys for input data and simulation models to design sustainable management systems and establish soil standards for managing soil quality.

Basic Soil Quality Indicators

A system to assess the health or quality of soil can be likened to a medical examination for humans (Larson & Pierce, 1991). In a medical exam, the physician takes certain key measurements such as temperature, blood pressure, pulse rate, and perhaps certain blood chemistries as basic indicators of body system function. If these basic health indicators are outside the commonly accepted ranges, more specific tests may be conducted to help identify the cause of the problem and find a solution. For example, excessively high blood pressure may indicate a potential for system failure (death) through stroke or cardiac arrest. The problem of high blood pressure may result from the lifestyle of the individual due to improper diet or high stress level. To assess a dietary cause for high blood pressure, the physician may request a secondary blood chemistry test for cholesterol, electrolytes, etc. Assessment of stress level as a causative factor for high blood pressure is less straightforward and generally involves implementing some change in lifestyle followed by periodic monitoring of blood pressure to assess the effect of change. This is a good example of using a basic indicator both to identify a problem and to monitor the effects of management on the health of a system.

A set of basic indicators of soil health or quality have not been previously defined, largely due to the difficulty in defining and identifying what soil quality represents and how it can be measured. The need for basic indicators of soil quality is indicated by the question commonly posed by farmers, reserachers, and conservationists: "What measurements should I make to evaluate the effects of management on soil function now and in the future?'' Our ability to identify basic soil properties that serve as indicators of soil quality is complicated by the many physical, chemical, and biological factors involved and their varying interactions in time, space, and intensity. However, one place to start is by identifying a basic list of measurable soil properties that define the major processes functioning in soil and ensure that the measurements we are making reflect conditions as they exist in the field. To be practical for use by scientists, farmers, extension workers, conservationists, ecologists, and policymakers over a range of ecological and socioeconomic situations, the set of basic soil quality indicators should meet the following suitability criteria:

1. Encompass ecosystem processes and relate to process oriented modeling.
2. Integrate soil physical, chemical, and biological properties and processes.
3. Be accessible to many users and applicable to field conditions
4. Be sensitive to variations in management and climate
5. Where possible, be components of existing soil data bases.

It is essential that basic soil quality indicators relate to ecosystem functions such as C and N cycling (Visser & Parkinson, 1992) and are components of driving variables for process-oriented models that emulate the ways ecosystems function. Evaluating the diverse effects of climate and management on soil function requires integration of basic physical, chemical, and biological indicators. *Too often scientists confine their interests and efforts to the discipline with which they are most familiar.* Microbiologists may confine their studies to microbial populations in the soil with little or no regard for soil physical or chemical characteristics that define the limits of microbial activity or that of other life forms. The approach to defining soil quality indicators must be holistic and not reductionistic. The indicators chosen must be accessible, or measurable, by as many people as possible and not limited to a select cadre of research scientists. Some indicators must be measured in the field to relate laboratory measurements to field conditions. One example is soil bulk density. Laboratory analyses for soil organic matter, N contents, etc. are commonly expressed on a gravimetric, or weight basis. These results are used to evaluate the effects of field management on soil quality as related to changes in organic matter or N content in the field. However, comparisons should not be made using gravimetric data, because soils are commonly sampled as a volume of soil to a specific depth. The specific concentration of a component in that volume of soil can change as the soil sample is prepared for analysis in the lab due to sieving, mixing, and drying. Thus expression of analytical results as concentration on a weight (gravimetric) basis does not accurately reflect the actual content or concentration for the soil depth sampled in the field. Also, soil bulk density frequently varies with management, depth of sampling, and time of year; valid comparison of management systems at a point in time or the same management system across time cannot be made without first adjusting results to a volumetric basis. Consequently, the inclusion of bulk density in a set of basic soil indicators is critical to proper interpretation of the importance of change in magnitude in other chemical and biochemical soil components.

Basic soil quality indicators should also be sensitive to variations in management or climate. If indicators of soil quality are insensitive to changes in management and climate, they will be of little use in monitoring changes in soil quality and proposing management changes to enhance soil quality. On the other hand, indicators of long-term changes in soil quality should not be confounded by short-term changes caused by seasonal weather patterns.

Wherever possible, basic indicators of soil quality should be properties or attributes that already exist in soil data banks. But we are faced with a formidable task in defining soil quality and the effects of management on sustainability. Basic indicators of soil quality will need to be compared with standards for a range of soils, climates, and management situations. Available data bases will be invaluable to formulation of standards, critical values, and thresholds for soil quality indicators.

A proposed set of basic soil quality indicators that meet many of the above-mentioned suitability criteria are given in Table 1-1. It should be emphasized that this is a set of *basic* indicators for initial characterization of soil quality. Other secondary measurements will likely be needed as dictated by existing data banks and specific climatic, geographic, and socioeconomic conditions or as indicated by assessment of basic indicators. Important soil quality indicators, not included in Table 1-1, are soil cation exchange capacity (CEC) and aggregate stability. However, CEC can be estimated from soil organic matter level, pH, and clay content. Inferences about soil aggregation can be made from organic matter content, soil infiltration rate after wetting, and soil bulk density (after wetting), which are included as basic indicators. However, a systematic cataloging of ranges and threshold values for these indicators will be needed to interpret the significance of changes in soil function and biological activity.

Larson and Pierce (1991) proposed that a minimum data set (MDS) be adopted for assessing the health of world soils and that standardized methodologies and procedures be established to assess changes in soil quality. They gave an example of a MDS of selected soil attributes for use in monitoring soil quality and changes with time (Table 1-2). Their choice of soil attributes for inclusion in the MDS was dictated by: (i) the need for selecting attributes sensitive to management and for which changes could be detected in a relatively short time, (ii) attributes for which measurement methodologies or data sets are accessible to most people, and (iii) attributes for which pedotransfer functions (Bouma, 1989) can be defined to interrelate soil properties and detect levels of soil quality. Many of the basic soil quality indicators given in Table 1-1 were taken from the MSD (Table 1-2) proposed by Larson and Pierce (1991).

The importance of standardized sampling methodologies and threshold values for interpretation of soil quality indicators cannot be over emphasized. The data presented in Table 1-3 represent measurements of soil quality indicators on soil samples from three farms in western Nebraska. The

Table 1-1. Proposed soil physical, chemical, and biological characteristics to be included as basic indicators of soil quality.

Soil characteristic	Methodology	Reference for methodology or interpretation, comments
Physical		
Soil texture	Hydrometer method	Gee & Bauder, 1986
Depth of soil and rooting	Soil coring or excavation	Taylor & Terrell, 1982
Soil bulk density and infiltration†	Field determined using infiltration rings	Blake & Hartge, 1986
Water holding capacity†	Field determined after irrigation of rings	Cassel & Nielsen, 1986
Water retention characteristics	Water content at 33 and 1500 kPa tension	Klute, 1986
Water content†	Gravimetric analysis; wt. loss, 24 h at 105°C	Sampled in field before and after irrigation
Soil temperature†	Dial thermometer or hand temperature probe	Measured at 4-cm soil depth
Chemical		
Total organic C and N	Wet or dry combustion, *volumetric basis*‡	Nelson & Sommers, 1982; Schulte, 1988
pH	Field or lab determined, pocket pH meter	Eckert, 1988; 1:1 soil/water mixture
Electrical conductivity	Field or lab, pocket conductivity meter	Dahnke & Whitney, 1988; 1:1 soil/water
Mineral N (NH_4 and NO_3), P, and K	Field or lab analysis, *volumetric basis*	Gelderman & Fixen, 1988; Knudsen & Beegle, 1988; $2\,M$ KCl extract for NH_4 and NO_3
Biological		
Microbial biomass C and N	Chloroform fumigation/incubation, *volumetric basis*	Parkinson & Paul, 1982
Potentially mineralizable N	Anaerobic incubation, *volumetric basis*	Keeney, 1982
Soil respiration†	Field measured using covered infiltration rings, lab measured in biomass assay	Anderson, 1982; CO_2-specific gas analysis tubes (Draeger)
Biomass C/Total org. C ratio	Calculated from other measures	Estimate of ecosystem stability; Visser & Parkinson, 1992; Chapt. 5, this book
Respiration/biomass ratio	Calculated from other measures	Visser & Parkinson, 1992; Chapt. 5, this book

† Measurements taken simultaneously in field for varying management conditions, landscape locations, and time of year.
‡ Gravimetric results must be adjusted to volumetric basis using field measured soil bulk density for meaningful interpretations.

Table 1-2. Soil attributes and methodologies for measurement to be included in a minimum data set (MDS) for monitoring soil quality (after Larson & Pierce, 1991).

Soil attribute	Methodology
Nutrient availability	Analytical soil test
Total organic C	Dry or wet combustion
Labile organic C	NH_4-N release from hot KCl digest
Particle size	Pipette or hydrometer methods
Plant-available water capacity	Best determined in the field or from water desorption curve
Soil structure, form	Bulk density from intact soil cores and field measured permeability or K_{sat}
Soil strength	Bulk density or penetration resistance
Maximum rooting depth	Crop specific, depth of roots or standard
pH	Glass/calomel electrode, pH meter
Electrical conductivity	Conductivity meter

producer on one of these farms used an innovative tillage management system that provided raised beds allowing the soil to warm faster in the spring, drain better, and form large aggregates (clods) to protect the soil from wind and water erosion. The farmer interpreted the 1.4-fold higher respiration, 5- to 15-fold greater total microbial biomass, and increased fungal biomass of his soil as compared with his neighbors soils as indicative of better *quality*. However, two issues related to interpretation of these measurements need further discussion. The first issue involves interpretation of laboratory results in relation to actual microbial processes in the field. The laboratory findings represent only potential differences, since the respiration measurements were conducted under conditions of optimal temperature and moisture, and results were not adjusted for soil bulk densities in the field at time of sampling. Actual differences could be 20 to 30% lower after adjusting for soil bulk density, since tilled areas would have bulk densities as low as 1.0 g cm^{-3} and nontilled areas as high as 1.2 to 1.3 g cm^{-3}. Secondly, without reference guidelines it is difficult to interpret the relevance of these measurements to soil quality. For example, the higher respiration and biomass of bedded ridges could reflect enhanced nutrient cycling and soil aggregation, both of which are considered positive soil quality attributes. However, a respiration rate of 84 mg C kg soil^{-1} d^{-1} might also represent depletion of soil C pools and soil organic matter. Assuming optimal conditions for soil respiration might

Table 1-3. Respiration and microbial biomass of 0- to 15-cm surface soils sampled from western Nebraska wheat fields in February 1991 (Parkin, 1991, unpublished data).

Area sampled	Soil respiration	Total microbial biomass	Fungal biomass[†]
	mg C kg^{-1} d^{-1}	mg C kg^{-1}	mg ergosterol kg^{-1}
Chemical–fallow farm	59	48	1.4
Conventional farm	54	18	0.9
Raised bed farm			
Ridge	84	262	2.4
Furrow	64	95	1.0

† Ergosterol as an estimate of fungal biomass (see Chapt. 16, this book).

occur for a 30-d period each year, the C loss for a 15-cm depth of soil in the bedded ridge would equal 3000 kg C yr^{-1}. This is three times the wheat (*Triticum aestivum* L.) residue C returned to this area (1000 kg C ha^{-1} yr^{-1}) and suggests a depletion of soil C reserves and long-term soil stability. From this standpoint, the high respiration rate observed would be undesirable and could detract from future soil quality. Clearly, additional guidelines for interpretation of soil quality indicators are badly needed.

Dynamic changes in soil quality indices and the many components of quality mandate temporal measurements and an evaluation approach that weighs the importance of various issues. As illustrated by the data presented in Table 1–4, evaluation of soil quality indicators varies during the growing season. The overall significance of soil quality indices such as microbial biomass N, potentially mineralizable N, and soil NO_3 levels with respect to crop productivity and environmental quality varies with time of year and agricultural management practice. For example, higher levels of microbial biomass and potentially mineralizable N, and lower levels of NO_3-N, in early spring (10 April) for alternative as compared with conventional management represent enhanced environmental quality due to lower potential N leaching losses during the nongrowing season but a potential limitation to corn (*Zea mays* L.) productivity. Decreases in biological N pools and corresponding increases in soil NO_3-N with alternative management during the growing season resulted in optimal final corn yield. These data illustrate the need to balance considerations for environmental quality with those for sustainable crop production. The alternative management system successfully reduced the inputs of synthetic chemicals, reduced the potential off-season losses of N, and maintained corn productivity by synchronizing the release of plant-available NO_3 from soil organic N sources with maximum plant need during the growing season. It is important to note, however, that the relative

Table 1–4. Management and cover crop effects on N pools in surface soil (0–30 cm) and related changes during the corn growing season in Pennsylvania (after Doran & Smith, 1991).

Management date	Microbial biomass N	Potentially mineralizable N	Nitrate N
		kg N ha^{-1}	
Alternative (corn–clover–winter wheat–soybean rotation + vetch cover crop)[†]			
10 April	121	1260	9
15 May[‡]	113	1260	39
12 June	75	1180	142
14 July	122	1220	99
Conventional (corn–soybean rotation with herbicides and fertilizer)			
10 April	92	990	42
15 May	56	950	56
12 June[§]	64	990	83
14 July	103	1020	106

† Corn, *Zea mays* L.; clover, *Trifolium* sp.; winter wheat, *Triticum aestivum* L.; soybean, *Glycine max* (L.) Merr.; vetch, *Vicia villosa* sp.
‡ Hairy vetch cover crop (182 kg N ha^{-1}) plowed into soil on 8 May.
§ 112 kg N ha^{-1} of ammonium nitrate fertilizer sidedressed on 17 June.

importance of soil quality issues and interpretation of soil quality issues varied with time of year.

Soil Quality Index—One Approach

There is general consensus that soil quality encompasses three broad issues; (i) plant and biological productivity, (ii) environmental quality, and (iii) human and animal health (Parr et al., 1992; Granatstein & Bezdicek, 1992; Arshad & Coen, 1992; Hornick, 1992). Therefore, any protocol designed to determine soil quality must provide an assessment of the function of soil with regard to these three issues. To effectively do this, the soil quality assessment must incorporate specific performance criteria for each of the three elements listed above, and it must be structured in such a way as to allow for quantitative evaluation and unambiguous interpretation.

Soil quality has been described as an inherent attribute of soil that may be inferred from its specific characteristics such as those presented in Table 1–2. However, measurement of the suite of properties listed in Table 1–2 will yield little insight into soil quality without specific criteria or guidelines for interpretation.

Two different approaches have been proposed for establishing reference criteria for assessing the quality of soil: (i) conditions of the native soil or (ii) conditions that maximize production and environmental performance (Granatstein & Bezdicek, 1992). For agricultural systems that are intensively managed we have adopted the latter approach and present below a framework for the evaluation of soil quality based on the function of soil.

A performance-based index of soil quality must provide an evaluation of soil function with regard to the major issues of (i) sustainable production, (ii) environmental quality, and (iii) human and animal health. To facilitate the development of specific performance criteria, we recommend that these three issues be further defined. *Sustainable production* can be defined in terms of plant production and resistance to erosion. *Environmental quality* can be defined in terms of groundwater quality, surface water quality, and air quality. *Human and animal health* can be defined in terms of food quality, which encompasses safety, and nutritional composition. Thus, we propose the following index of soil quality as a function of six specific soil quality elements (Eq. [1]):

$$SQ = f(SQ_{E1}, SQ_{E2}, SQ_{E3}, SQ_{E4}, SQ_{E5}, SQ_{E6}) \qquad [1]$$

where the specific soil quality elements (SQ_{Ei}) are defined as follows:
SQ_{E1} = food and fiber production
SQ_{E2} = erosivity
SQ_{E3} = groundwater quality
SQ_{E4} = surface water quality
SQ_{E5} = air quality
SQ_{E6} = food quality

The advantage of this approach is that the functions of soil can be assessed based on specific performance criteria established for each element, for a given ecosystem. For example, yield goals for crop production (SQ_{E1}); limits for erosion losses (SQ_{E2}); concentration limits for chemical leaching from the rooting zone (SQ_{E3}); nutrient, chemical, and sediment loading limits to adjacent surface water systems (SQ_{E4}); production and uptake rates for trace gases that contribute to O_3 destruction or the greenhouse effect (SQ_{E5}); and nutritional composition and chemical residue of food (SQ_{E6}).

At this time there is not sufficient information to identify, with certainty, the optimum functional relationship used to combine the different soil quality elements shown in Eq. [1]; however, one possibility is a simple multiplicative function (Eq. [2]).

$$SQ = (K_1 SQ_{E1})(K_2 SQ_{E2})(K_3 SQ_{E3})(K_4 SQ_{E4})(K_5 SQ_{E5})(K_6 SQ_{E6}) \qquad [2]$$

where K = weighting coefficients.

In a manner analogous to the soil tilth index of Singh et al. (1990), weighting factors are assigned to each soil quality element, with the relative weights of these coefficients being determined by geographical considerations, societal concerns, and economic constraints. For example, in a given region, food production may be the primary concern, and elements such as air quality may be of secondary importance. If such were the case, SQ_{E1} would be weighted more heavily than SQ_{E5}. Thus, this framework has an inherent flexibility in that the precise functional relationship for a given region, or a given field, is determined by the intended use of that area or site, as dictated by geographical and climatic constraints as well as socioeconomic concerns.

Implementation of the Index

It is proposed that each soil quality element in this index be evaluated with regard to five specific soil functions, which define the capacity of soil to (i) provide a medium for plant growth and biological activity, (ii) regulate and partition water flow through the environment, and (iii) serve as an effective environmental filter (Larson & Pierce, 1991). These specific soil function factors are:

SF_1 = ability to hold, accept, and release water to plants, streams, and subsoil (water flux)
SF_2 = ability to hold, accept, and release nutrients and other chemicals (nutrient and chemical fluxes)
SF_3 = promote and sustain root growth
SF_4 = maintain suitable soil biotic habitat
SF_5 = respond to management and resist degradation

The evaluation of each soil quality element will take the form of a functional relationship that describes how the five soil functions listed above impact each of the different soil quality elements.

$$SQ_{E1} = f(SF_1, SF_2, SF_3, SF_4, SF_5)$$

$$SQ_{E2} = f(SF_1, SF_2, SF_3, SF_4, SF_5)$$

$$SQ_{E3} = f(SF_1,SF_2,SF_3,SF_4,SF_5)$$

$$SQ_{E4} = f(SF_1,SF_2,SF_3,SF_4,SF_5)$$

$$SQ_{E5} = f(SF_1,SF_2,SF_3,SF_4,SF_5)$$

$$SQ_{E6} = f(SF_1,SF_2,SF_3,SF_4,SF_5)$$

It is apparent that with this approach, an assessment of soil quality essentially requires the evaluation of six separate functions. The rationale for this approach is necessitated by the fact that soil functions in a duplicitous manner. For example, the attributes at a given site, such as the presence of a clay pan, may serve to retard the leaching of chemicals from the rooting zone, which could be viewed as beneficial from an environmental quality perspective, yet the same clay pan might also restrict the development of crop rooting systems, and thus is detrimental from a productivity standpoint. Thus, different mathematical relationships relating the soil functions to each soil quality element must be developed for each soil quality element.

The next step in the process is to develop mathematical expressions that relate the five soil function components listed above (SF_i) to the set of basic soil attributes or processes shown in Table 1-2. It is recognized that, for each soil quality element, the mathematical expression for a given soil function component SF_i will take a different form. An example of this approach is given where SF_1 is related to soil quality elements SQ_{E1}, SQ_{E2}, and SQ_{E3}. For this example SF_1, the ability to hold, accept, and release water, is represented as a function of infiltration rate and water retention.

$$SF_1 = f(\text{Infiltration rate, Water retention})$$

Theoretical examples of how soil infiltration rate and water retention might be mathematically related to these soil quality elements are presented in Fig. 1-2. It is noted that the mathematical function relating infiltration and water retention to soil quality takes a different form for each of the three soil quality elements. Increased infiltration and water retention result in a higher-quality element rating for crop production (SQ_{E1}), but a lower rating for groundwater quality. Erosivity is independent of water retention in the soil profile and is largely influenced by infiltration. This is the dilemma we are commonly faced with in assessing soil quality; soil properties that enhance one soil quality element may detract from another. Depending on the particular site being assessed, weighting factors based on the intended use of the site will have been predetermined (based on socioeconomic factors) to emphasize the importance of one element over another.

For some soil functions, separate evaluations may be required. For SF_2 (ability to accept, hold, and release nutrients and other chemicals), separate evaluations are required for individual nutrients and chemicals such as N, P, K, heavy metals, and pesticides. It should be noted that these mathematical expressions can be developed to account for regional variations induced by specific cropping systems, geographical location, and climate.

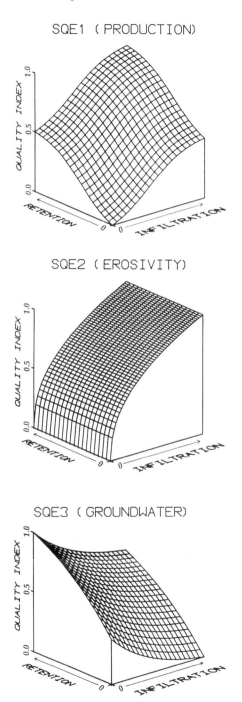

Fig. 1-2. Theoretical examples of how soil attributes can be mathematically related to soil quality elements.

To be of practical use, this approach of defining soil quality based on function, soil function must be related to measurable soil attributes. In the example illustrated in Fig. 1-2, infiltration and water retention could be measured directly or predicted from basic soil attributes (bulk density, organic matter, conductivity) in a manner similar to the pedotransfer functions described by Larson and Pierce (1991; Chapt. 3, this book, 1994). The development of relationships between soil attributes and soil functions may be a monumental task. However, algorithms in existing simulation models (e.g., NLEAP, EPIC, CREAMS, GLEAMS, WEPP) may serve as a useful starting point.

The purpose of this proposed approach to a soil quality index is not to replace previous work in the area of soil quality assessment; rather, it is intended to complement past work by presenting a more clearly defined framework for the development of mathematical relationships, driven by basic soil attributes, to define soil quality. The proposed approach encompasses the flexibility required to be effective over a range of ecological and socioeconomic situations.

CONCLUSION AND NEEDS

There is great need both to determine the status of and to enhance soil resources. Assessment and monitoring of soil quality must also provide opportunity to evaluate and redesign soil and land management systems for sustainability. We need standards of soil quality to determine what is good or bad and to find out if soil management systems are functioning at acceptable levels of performance. We see the following areas as research needs critically important to assessment and enhancement of soil quality. Although we have discussed many of these needs in this chapter, most have been alluded to by authors of other chapters in this soil quality publication.

1. For valid comparison of soil quality across variations in climate, soil, and management, we must establish reference guidelines and thresholds for soil quality indicators that enable interpreting relationships between measured soil attributes and soil function. This will require establishing appropriate scales of time and space for soil quality assessment and standardizing methods and protocols for sampling, processing, and analysis.

2. Develop a practical index for on-site assessment of soil quality by farmers, researchers, extension personnel, and environmental monitorers that can also be used by resource managers and policymakers.

3. Determine the effects of soil quality on plant growth, nutritional composition, and related animal and human health. Provide standards for food quality in terms of specified levels of key nutrients. Identify indicators of soil quality that can be related to food quality and human health.

4. Assess the current status of the biological, physical, and chemical properties of benchmark soils in major management groups throughout the USA. This research would extend information currently being collected by USDA-SCS and SAES soil surveys and genesis and morphology surveys. This assessment should be coordinated with the USEPA sponsored EMAP (Environmental Monitoring and Assessment Program) to which considerable resources have already been committed.

5. Determine the effects of current cropping and management systems and proposed sustainable systems on organic matter levels and other soil properties. Assess the relative effects of increasing or decreasing soil organic matter levels in managed soil ecosystems on atmospheric CO_2 and other environmentally important gases and on global climate change predictions.

6. Develop methods and criteria for socioeconomic assessment of soil and land management systems. Estimate economic impacts of improving *soil quality* including increased productivity, increased pollution abatement, decreased sedimentation, increased nutrient use efficiency, and decreased use of energy in crop production. Evaluate how soil quality assessment can be used to estimate economic return from public investment in conservation practices such as the Conservation Reserve Program.

7. Identify biological indicators of soil quality that assess soil biological diversity and food-chain levels in soil as related to soil biological health and nutrient cycling.

8. Develop precision farming techniques for quality enhancement of soils. Establish management practices necessary to attain the biological diversity and food-chain levels for acceptable sustainability.

9. Develop sensors and sensing technology for static and real-time measurement of key variables that define soil quality or are indicative of changes in soil quality. Develop remote sensing capabilities for large-scale assessment of soil quality changes over time.

REFERENCES

Anderson, J.P.E. 1982. Soil respiration. *In* A.L. Page et al. (ed.) Methods of soil analysis. Part 2. 2nd ed. Agron. Monogr. 9. ASA and SSSA, Madison, WI.

Arshad, M.A., and G.M. Coen. 1992. Characterization of soil quality: Physical and chemical criteria. Am. J. Altern. Agric. 7:5–12.

Bhagat, S.P. 1990. Creation in crisis. Brethren Press, Elgin, IL.

Blake, G.R., and K.H. Hartge. 1986. Bulk density. p. 363–375. *In* A. Klute (ed.) Methods of soil analysis. Part 1. 2nd ed. Agron. Monogr. 9. ASA and SSSA, Madison, WI.

Bouma, J. 1989. Using soil survey data for quantitative land evaluations. Adv. Soil Sci. 9:177–213.

Cassel, D.K., and D.R. Nielsen. 1986. Field capacity and available water capacity. p. 906–908. *In* A. Klute (ed.) Methods of soil analysis. Part 1. 2nd ed. Agron. Monogr. 9. ASA and SSSA, Madison, WI.

Council for Agricultural Science and Technology. 1992. Preparing U.S. agriculture for global climate change. Task force rcp. 119. CAST, Ames, IA.

Dahnke, W.C., and D.A. Whitney. 1988. Measurement of soil salinity. p. 32–34. *In* Recommended chemical soil test procedures for the North Central Region. North Central Regional Publ. 221. North Dakota Agric. Exp. Stn. Bull. 499.

Doran, J.W., and M.S. Smith. 1991. Role of cover crops in nitrogen cycling. p. 85–90. *In* W.L. Hargrove (ed.) Cover crops for clean water. Proc. of an international conference. Jackson, TN. 9–11 Apr. 1991. Soil and Water Conserv. Soc., Ankeny, IA.

Eckert, D.J. 1988. Recommended pH and lime requirement tests. p. 6–8. *In* Recommended chemical soil test procedures for the North Central Region. North Central Regional Publ. 221. North Dakota Agric. Exp. Stn. Bull. 499.

Gee, G.W., and J.W. Bauder. 1986. Particle-size analysis. p. 383–409. *In* A. Klute (ed.) Methods of soil analysis. Part 1. 2nd ed. Agron. Monogr. 9. ASA and SSSA, Madison, WI.

Gelderman, R.H., and P.E. Fixen. 1988. Recommended nitrate-N tests. p. 10–12. *In* Recommended chemical soil test procedures for the North Central Region. North Central Regional Publ. 221. North Dakota Agric. Exp. Stn. Bull. 499.

Granatstein, D., and D.F. Bezdicek. 1992. The need for a soil quality index: Local and regional perspectives. Am. J. Altern. Agric. 17:12–16.

Grove, P.B. (ed.). 1986. Websters Third New International Dictionary. Merriam-Webster Inc. Publishers, Springfield, MA.

Haberern, J. 1992. Viewpoint: A soil health index. J. Soil Water Conserv. 47:6.

Hornick, S.B. 1992. Factors affecting the nutritional quality of crops. Am. J. Altern. Agric. 7:63–68.

Houghton, R.A., J.E. Hobbie, J.M. Melillo, B. Moore, B.J. Peterson, G.R. Shaver, and G.M. Woodwell. 1983. Changes in the carbon content of terrestrial biota and soils between 1860 and 1980: A new release of CO_2 to the atmosphere. Ecol. Monogr. 53:235–262.

Jenny, H. 1980. The soil resource: Origin and behavior. Ecol. Studies 37. Springer-Verlag, New York.

Keeney, D.R. 1982. Nitrogen—availability indices. p. 711–733. *In* A. Klute (ed.) Methods of soil analysis. Part 1. 2nd ed. Agron. Monogr. 9. ASA and SSSA, Madison, WI.

Klute, A. 1986. Water retention: Laboratory methods. p. 635–662. *In* A. Klute (ed.) Methods of soil analysis. Part 1. 2nd ed. Agron. Monogr. 9. ASA and SSSA, Madison, WI.

Knudsen, D., and D. Beegle. 1988. Recommended phosphorus tests. p. 12–14. *In* Recommended chemical soil test procedures for the North Central Region. North Central Regional Publ. 221. North Dakota Agric. Exp. Stn. Bull. 499.

Larson, W.E., and F.J. Pierce. 1991. Conservation and enhancement of soil quality. *In* Evaluation for sustainable land management in the developing world. Vol. 2. IBSRAM Proc. 12 (2). Bangkok, Thailand. Int. Board for Soil Res. and Management.

Mosier, A.R., D. Schimel, D. Valentine, K. Bronson, and W. Parton. 1991. Methane and nitrous oxided fluxes in native, fertilized and cultivated grasslands. Nature (London) 350:330–332.

Nelson, D.W., and L.E. Sommers. 1982. Total carbon, organic carbon, and organic matter. p. 539–579. *In* A.L. Page (ed.) Methods of soil analysis. Part 2. 2nd ed. Agron. Monogr. 9. ASA and SSSA, Madison, WI.

Papendick, R.I., and J.F. Parr. 1992. Soil quality—the key to a sustainable agriculture. Am. J. Altern. Agric. 7:2–3.

Parkinson, D., and E.A. Paul. 1982. Microbial biomass. p. 821–830. *In* A.L. Page (ed.) Methods of soil analysis. Part 2. 2nd ed. Agron. Monogr. 9. ASA and SSSA, Madison, WI.

Parr, J.F,. R.I. Papendick, S.B. Hornick, and R.E. Meyer. 1992. Soil quality: Attributes and relationship to alternative and sustainable agriculture. Am. J. Altern. Agric. 7:5–11.

Pierce, F.J., and W.E. Larson. 1993. Developing criteria to evaluate sustainable land management. p. 7–14. *In* J.M. Kimble (ed.) Proc. of the 8th Int. Soil Management Workshop; Utilization of Soil Survey Information for Sustainable Land Use. May 1993. USDA-SCS, National Soil Surv. Center, Lincoln, NE.

Power, J.F., and R.J.K. Myers. 1989. The maintenance or improvement of farming systems in North America and Australia. p. 273–292. *In* J.W.B. Stewart (ed.) Soil quality in semi-arid agriculture. Proc. of an Int. Conf. sponsored by the Canadian Int. Development Agency, Saskatoon, Saskatchewan, Canada. 11–16 June 1989. Saskatchewan Inst. of Pedology, Univ. of Saskatchewan, Saskatoon, Canada.

Rodale Institute. 1991. Conference report and abstracts, Int. Conf. on the Assessment and Monitoring of Soil Quality. Emmaus, PA. 11–13 July 1991. Rodale Press, Emmaus, PA.

Sagan, C. 1992. To avert a common danger. Parade Magazine, March 1, p. 10–14.

Sanders, D.W. 1992. International activities in assessing and monitoring soil degradation. Am. J. Altern. Agric. 7:17–24.

Schulte, E.E. 1988. Recommended soil organic matter tests. p. 29–32. *In* Recommended chemical soil test procedures for the North Central Region. North Central Regional Publ. 221. North Dakota Agric. Exp. Stn. Bull. 499.

Singh, K.K., T.S. Colvin, D.C. Erbach, and A.Q. Mughal. 1990. Tilth index: An approach towards soil condition quantification. Meeting Pap. 90-1040. Am. Soc. of Agric. Eng., St. Joseph, MI.

Soil Science Society of America. 1987. Glossary of soil science terms. SSSA, Madison, WI.

Taylor, H.M., and E.E. Terrell. 1982. p. 185–200. *In* M. Rechcigl, Jr. (ed.) CRC handbook of agricultural productivity. Vol. 1. CRC Press, Boca Raton, FL.

Visser, S., and D. Parkinson. 1992. Soil biological criteria as indicators of soil quality: Soil microorganisms. Am. J. Altern. Agric. 7:33–37.

World Resources Institute. 1992. New York Times, March 3.

2 Descriptive Aspects of Soil Quality/Health

R. F. Harris

University of Wisconsin
Madison, Wisconsin

D. F. Bezdicek

Washington State University
Pullman, Washington

The terms *soil quality* and *soil health* are currently used interchangeably in the scientific literature and popular press. In general, scientists prefer soil quality and farmers prefer soil health. Alternative agricultural institutes such as Rodale (PA) use soil quality and soil health without qualification (Haberern, 1992a,b). We favor using the joint term *soil quality/health* in the interest of promoting communication, knowledge sharing, and developing an understanding of the language and methods used to manage soil quality/health by farmers and scientists (Harris, 1992; Harris et al., 1992).

The need to develop methodology for characterizing soil quality/health is gaining increasing global recognition. At the international level, recent developments in soil quality/health include: (i) an international conference on "Assessment amd Monitoring of Soil Quality" held in 1991 at the Rodale Institute in Pennsylvania (Papendick & Parr, 1992); (ii) an "International Workshop on Evaluation for Sustainable Land Management in the Developing World" held in Thailand in 1991, had as a topic highlight a paper on "Conservation and Enhancement of Soil Quality" (Larson & Pierce, 1991); (iii) the 7th International Soil Conservation Organization Conference held in Australia in 1992, with "People Protecting Their Land" as its theme, had a lead-off session focusing on the question, "How do we assess the health or condition of the land?"; (iv) the 1992 European Conference on Integrated Research for Soil Protection held in the Netherlands had a major session on "Assessment of Soil Quality and Soil Vulnerability"; and (v) The CSSS/CLRA/ISSS/ISATCC International Conference on "Environmental Soil Science" had four invited papers on "Criteria, Indicators and Applications of Soil Quality." Recent national-level developments in soil quality/health include: (i) a university/federal research committee (NCR-59), based in the midwest but including nationwide representation, was reoriented in 1991 to focus on development and validation of methods for characterizing soil quality, particularly the biotic component; and (ii) the Soil Science Society of America held a symposium titled "Defining Soil Quality for a Sustainable Environment" at its 1992 annual meeting, on which this special publication is based.

Diagnostic properties characterizing the quality/health of soil, plant, animal–human, water or air target systems, subdivide into two major components, descriptive and analytical. Properties under the descriptive component use words as descriptors, and are thus inherently qualitative and subjective; descriptive data are generally considered "soft" and of limited value by natural scientists and other technical experts. Properties under the analytical component use units as descriptors, and are thus quantitative; analytical data are considered "hard" and thus more acceptable by technical experts. User groups tend to communicate with descriptive properties, whereas technical experts emphasize analytical methods for unambiguous, quantitative identification and measurement of diagnostic properties. Communication between user groups and technical experts must involve a combination of descriptive and analytical properties, with relative emphasis being a function of the target system. Validation of assessment methodology which predicts the capability of a system to function requires extensive correlative evaluation and documentation of diagnostic properties and target system performance.

This chapter evaluates the role of descriptive soil quality/health information in the development and application of technical and nontechnical tools for assessing and monitoring soil quality/health for user groups of diverse backgrounds and interests.

HISTORICAL

Since ancient times farmers have recognized and managed the quality/health of the soil, using descriptive terms to assess the condition of the soil and its performance as a sustainable crop production system. As reviewed by Karlen et al. (1990), Varro, a Roman scholar, noted that certain crops "when cut down and left, they improve the soil"; and Fream (1890) quoted 17th century Thomas Tusser "good tilth brings seeds, ill tilture weeds," and noted himself that a soil "open, free working, or in good heart makes us feel good about it" but if a soil is "hungry, stubborn, stiff, or unkind," we immediately perceive it as nonproductive.

With the advent of soil and crop science in the early 1900s, there has been a progressive shift from descriptive to analytical properties and an increasing reliance on scientists for assessment, monitoring, and management advice. The focus has been almost exclusively on the inorganic chemical (soil test) and its relationship to crop yield.

In recent years there has been a growing awareness by alternative farmers, environmentalists, and ecologists that soil quality/health involves more than inorganic chemical soil tests and crop yield. Most now view the soil as a delicately balanced ecosystem supporting a diverse web of organisms within and on its surface, and as a living buffer and filter for protecting the surface water, groundwater, and air.

The role of descriptive soil information is not commonly covered in the scientific literature dealing with soil quality/health characterization. However, Arshad and Coen (1992) recommend that qualitative (descriptive) information should form an essential part of soil quality monitoring programs. As they point out, visual, morphological observations can be used by both farmers and scientists in the field to recognize deteriorated soil quality caused by (i) loss of organic matter, reduced aggregation, low hydraulic conductivity, soil crusting, and surface sealing; (ii) water erosion, as indicated by rills, gulleys, stones on surface, exposed roots, and uneven topsoil; (iii) wind erosion, as indicated by ripple marks, sand against plant stems, and plant damage; (iv) salinization, as indicated by salt crust and salt-tolerant plants; (v) acidification and chemical degradation, as indicated by acid-tolerant plants and lack of fertilizer response; (vi) poor drainage and compact–hardpan structural deterioration, as indicated by standing water, and poor and patchy crop stand. In a more general sense, observational information on soil characteristics such as color, texture, structure, and consistence, has long been used as an integral part of soil survey procedures.

The concept of soil health is in many ways farmer- rather than institution-driven. Farmer concerns about soil health are rooted in observational field experiences, which translate into descriptive properties of soil and systems supported by or interfacing with soil. The potential contributions of farmer "indigenous knowledge" to conceptual, experimental, and management of soil quality/health, in both developed countries and developing countries (Pawluk et al., 1992), has yet to be fully explored.

WASHINGTON STATE FARMER PERCEPTIONS

As a first effort to investigate sustainable agriculture in the Palouse region of Washington and Idaho, 24 farmers were identified and selected on the basis of their reputed use of lower input and other alternative farming practices (Beus et al., 1990; Granatstein, 1990). The farmers fell roughly into three groups: (i) those who were using unique cropping rotations or systems that they felt helped them build soil, reduce erosion, and in some instances decrease input costs; (ii) those who were significantly cutting back on purchased inputs (i.e., fertilizers and pesticides), primarily through cautious experimentation; and (iii) those who were using alternative or biological products as a major part of their production system.

Most of the farmers placed considerable emphasis on soil quality, soil health, and soil organic matter. They felt that building a healthy soil was one of the keys to sustainable agriculture. By so doing, many believed that improvement in other areas, such as preventing plant diseases and reducing soil erosion, would naturally follow. Many identified improving soil tilth, enhancing the microbiological health of the soil, and increasing soil organic matter as primary goals of their production system. A commonly voiced frustration among many of the farmers was their perception that industry and university research did not pay nearly enough attention to these issues. The farmers expressed their concern that research is becoming too narrowly focused and thus is not able to investigate the holistic relationship of soil health to plant health, productivity, and environmental protection.

The farmers recognized soil organic matter as a key ingredient of soil quality, and observed that increased soil organic matter improved soil moisture storage, reduced erosion, and made tillage easier. They were also keenly aware of the importance of good soil structure for water infiltration and storage, root health, reduced draft requirement, and erosion control. For example, in a comment on the beneficial effects of grass on soil tilth, a farmer noted, "You can definitely see that there is an advantage...You can see it as you work the ground. It's very obvious to the eye in the way the tractor pulls and things like that." Another farmer observed that, after changing management practice, his soil is now "like walking on a carpet instead of a highway. Fuel costs are way down and our wear and tear on our machinery is way down. Our ground just farms easier than it used to." One farmer noticed that in response to changing management, his ground was "less shiny than the neighbors, with no standing water."

One indicator used by several farmers to recognize a microbiologically healthy soil was the visually observable rate of incorporated straw decomposition. In a comment on this issue, a farmer noted "...in our field, I plowed down over 100 bushel straw and next spring I went in with the cultivator and it was black. I plowed another place, and I was turning up years of straw, that gray stuff. It hadn't decomposed. Some ground is just healthier than others." Another indicator of soil health used by the farmers was the observable presence of earthworms. The strong dependence of farmers on descriptive indices of soil quality/health is illustrated by the concluding comment on soil biology by Granatstein (1990): "the observations of these farmers are very qualitative, but illustrate their feeling that a biologically active soil is desirable."

WISCONSIN FARMER PERCEPTIONS

The Wisconsin Soil Quality/Health program was initiated in response to farmer concern about the need for university research on soil biological quality and soil health. A major initial objective was to find out the basis of farmer concerns in this area to provide a foundation for development of analytical soil tests.

Conversations with farmers in the field during the spring of 1990 provided farmer perceptions of soil health and led to the development of a one-page questionnaire based on simple sensory (look, smell, feel) properties of soil associated with soil health by farmers. The questionnaire was distributed at the 1990 sustainable agriculture field day and at a state fair booth. The major rationale for the farmer descriptive information was to provide insight into a definition of soil quality/health that could be used as a basis

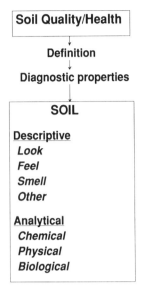

Fig. 2–1. Soil quality/health interpretive framework: 1990 field day.

for identification of a meaningful set of analytical properties. The approach may be summarized in the form of a first approximation interpretive framework that recognized (i) the presumed need for a reference definition of soil quality/health, and (ii) diagnostic soil quality/health properties based on look, smell, and feel descriptive properties of soil and chemical, physical, and biological analytical properties of soil (Fig. 2–1).

The results of the questionnaire led to a second approximation interpretive framework (Fig. 2–2) that again recognized the presumed need for a reference definition of soil quality/health, but expanded the diagnostic soil quality/health properties to include properties of plant and farm target systems as well as the soil target system (Fig. 2–2). In the summer of 1991, the Wisconsin Soil Quality/Health program conducted field interviews of 25 farmers in four regions across Wisconsin to gain their perceptions about healthy and unhealthy soils (Porter, 1991). The farmers represented five farming types: high-input conventional, low-input conventional, organic, biologic, and biodynamic. The interview was conducted outside, and farmers were asked to describe a field they thought was healthy or unhealthy. They were then prompted with 15 questions about what they noticed about soil health. The questions followed the second approximation interpretive framework, and accordingly focused on descriptive characteristics of the soil and interacting systems, including: soil appearance, smell, and structure; crop stress tolerance, yield, and quality; and farm health. Farmer perceptions of soils that they considered healthy typically included comments that such soils were deeper, darker, easier to plow (higher gear), worked up more easily in the spring, sponged up and held more water, dried out sooner, broke down corn (*Zea mays* L.) stalks more rapidly in the fall, had higher organic matter and

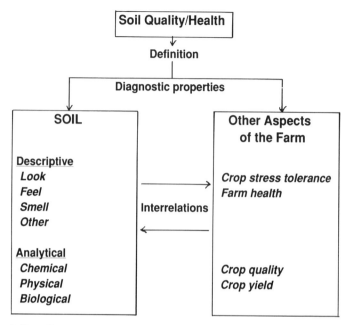

Fig. 2-2. Soil quality/health interpretive framework: 1991 field interview and 1992 regional focus meetings.

less erosion, had greater numbers and more species of worms, and had a recognizably sweet, fresh-air smell. Some farmers maintained that healthy soils had a higher calcium to magnesium ratio. Other farmers noted that crops grown on healthy soils required less nutrients, gave higher crop yields, tested higher in mineral and plant sugar (BRIX) levels, had a diverse rather than single weed population, had fewer disease and insect problems, and were more resistant to drought. Others said that their cows, when fed crops from healthy soils, produced more on less feed, and had noticeably lower veterinarian bills.

In late winter of 1992, four regional meetings were hosted by farmer members of the Soil Quality/Health program steering committee to provide a forum for discussion and feedback on the progress and aims of the soil quality/health project, and to obtain further insights into farmer perceptions of soil health (Andersen, 1992). These *town hall* meetings were open to everyone in the surrounding area, and were attended by a total of about 70 farmers, county agents, and the general public. Much of the meeting time was spent exploring, with appropriate prompts using the second approximation soil quality/health interpretive framework (Fig. 2-2) as a guide, how group members recognized healthy and unhealthy soils. Specific comments were very similar to those from the 1991 field interviews. A recurrent theme was "healthy soil, healthy plants, healthy animals, healthy people." A major conclusion from the meeting was that many farmers recognize that a healthy soil is based on more than soil properties, but is also based on the health or quality of the plants and animals supported by the soil and the water and air interfacing with the soil.

At a 1992 field day at the University of Wisconsin Arlington Agricultural Research Station, the Wisconsin Soil Quality/Health program explained to diverse audiences that we were determining the language and methods used by farmers and scientists to recognize and manage soil quality/health, with the ultimate goal of developing technical and nontechnical tools for assessment and tracking of the effect of management practices on soil quality/health. Without further elaboration, we asked the group to respond at will to the question, "How do you recognize a high-quality/healthy or low-quality/unhealthy soil?" We recorded the answers on blank flip charts, locating each answer according to whether it fit the descriptive vs. analytical component of the soil vs. nonsoil category. Headings were not identified to prevent an audience response bias. The answers confirmed the trends found earlier using the one-on-one unprompted and prompted interview approach, but provided some additional insights because of the nature of the questioning and the broader cross-section of interest groups. For example, one home gardener said that she recognized whether she had a *good* soil solely as a function of the ease of pulling weeds out of the ground. Many said that they associated healthy soil with soil smell, but had real problems defining in words what they meant. Some conventional farmers said that they depended on technical consultants to diagnose the quality of their soils using standard soil tests and recommend best management practices for maximized crop yield, but felt there were ill-defined issues of soil stewardship that also needed to be taken into account somehow. The dialogue with the tour groups was concluded by demonstrating to the groups that all their answers to the soil quality/health recognition question fit into the interpretive framework within the descriptive or analytical soil or nonsoil target system categories, and that all concepts of soil quality/health were interlinked and recognizable within the interpretive framework.

The farmer interviews, focus group meetings, field stop interactions, and follow-up interviews with participating farmers (Porter et al., 1992) provided an information base for the final soil quality/health interpretive framework (Fig. 2–3), discussed later.

TOOLS FOR SOIL QUALITY/HEALTH CHARACTERIZATION AND MANAGEMENT

Farmer–scientist partnership on soil quality/health is essential for the integration of farmer observational knowledge of soil quality/health with institutional analytical knowledge in development of soil quality/health assessment tools and management guidelines for economically and environmentally sustainable use of soil resources. Soil quality/health assessment provides a classic opportunity for productive and mutually beneficial interaction between natural and social scientists, an elusive and uncommon phenomenon because of peer recognition problems and related project fund-

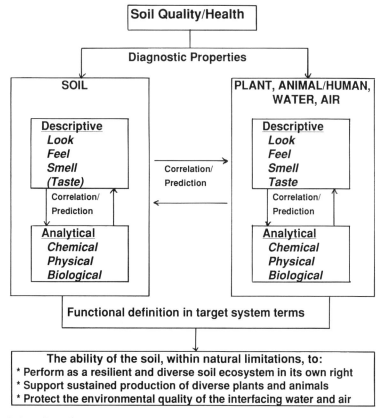

Fig. 2-3. Soil quality/health interpretive framework: current.

ing limitations (Heberlein, 1988). A set of conceptual, information gathering, and management tools for assessing soil quality/health has evolved from farmer/scientist partnership in Wisconsin and Washington. The following focuses on the function of these tools and the role of descriptive aspects of soil quality/health in them. Additional details are given in Garlynd et al. (Chapt. 10, this book).

Conceptual Tool for Soil Quality/Health Assessment

Interpretive Framework. An interpretive framework facilitates identification of the critical issues and major categories of information that need to be gathered for development of correlations needed to establish predictive capability. Our version (Fig. 2-3) recognizes that the target systems for soil quality/health characterization are the soil itself, and the plant, animal-human, water, and air systems supported by and interfacing with soil. For each target system, the descriptive components are grouped into the major sensory categories of look, feel, smell, and taste, and analytical components into classical chemical, physical, and biological properties. A data base of descriptive and analytical properties of the soil and nonsoil target systems is needed to establish correlations within and between target systems and achieve the ultimate objective of predicting performance as a function of soil properties. It follows that the definition of *soil quality/health* becomes the capability of soil (i) to function as a resilient and diverse ecosystem in its own right, (ii) to support sustained production of diverse plants and animals, and (iii) to protect the environmental quality of the interfacing water and air (Fig. 2-3).

The most important implication of the interpretive framework (Fig. 2-3) is that all experimental designs for soil quality/health work should take into account all target systems recognized as being an integral part of soil quality/health. These target systems are the soil itself and, just as important if not more important in many people's perceptions, the crops and animals-humans supported by soil, and the water and air interfacing with soil. The ultimate objective of soil quality/health work is to be able to predict from a knowledge of soil properties, the ability of the soil to carry out functions resulting in specific outcomes, expressed in terms of specific quality/health properties of the crop, animal–human, water, and air target systems. Attainment of such a predictive capability requires correlative work with a data base, including the properties to be predicted as well as those from which the prediction is to be made. For example, predictions of crop growth from soil inorganic fertility tests have required extensive field studies on correlative relationships between soil properties as the independent variable and crop yield as the dependent variable.

In practice, for most research groups, it will be logistically impossible to establish an experimental protocol to measure all the descriptive and particularly analytical diagnostic properties of the quality/health of the soil, crop, animal–human, water, and air target systems (Fig. 2-3). However, knowledge of the ideal allows for more informed decisions to be made and defended as to which properties are to be included and which are to be ignored while still achieving a viable experimental program. In practice, except for standard crop yield, the nonsoil analytical components of the crop, animal–human, water and air target systems will likely be the most difficult to obtain routinely. However, if a systematic protocol for the descriptive components of the quality/health of these systems were available, a farmer–scientist partnership analogous to a patient–doctor collaboration on descriptive monitoring of health signs would appear to be a feasible goal.

Information-Gathering Tools for Soil Quality/Health Assessment

The information-gathering tools for soil quality/health assessment should be internally consistent with the interpretive framework (Fig. 2-3).

Interview Guide. A comprehensive assessment of how diverse interest groups recognize and measure soil quality/health is needed so that indigenous descriptive knowledge of progressive farmers can be maximized to contribute to the information base of soil quality/health. This information can then be distilled into a protocol usable by farmers and scientists, independently and in partnership, for assessing and monitoring the effect of management on soil quality/health.

The Wisconsin interview guide (Chapt. 10, this book) consists of forms that recognize the interpretive framework categorization of diagnostic properties of soil quality/health into target system, descriptive and analytical components, and major categories of look, smell, feel and taste (descriptive), and chemical, physical, and biological (analytical) properties. Blank spaces are included for recording specific properties with qualifying word or unit descriptors used by respondents to recognize high quality/healthy and low quality/unhealthy soils. Additional forms allow recording of correlation perceptions and respondent background information. Implementation of the interview guide is patterned after the approach described earlier (see Wisconsin Farmer Perceptions section) for the 1992 field day at the University of Wisconsin Arlington Agricultural Research Station (Chapt. 10, this book).

Questionnaires. The questionnaires should follow the target system, descriptive and analytical component, and major category structure of the interpretive framework and interview guide, and results from application of the interview guide provide a basis for completion to the level of a minimum data set of specific descriptive and analytical diagnostic properties characterizing soil quality/health.

The Wisconsin Soil Quality/Health program currently uses two forms of the questionnaire (Chapt. 10, this book): (i) a general questionnaire for establishing the language used by diverse farmers, scientists, and other groups to recognize and manage soil quality/health, and (ii) a site-specific questionnaire for obtaining farmer input to the descriptive component of the correlative report card for a site identified for detailed sampling and correlative evaluation.

Correlative Report Card. The correlative report card is the vehicle used to record detailed descriptive and analytical data from experimental sites targeted for soil quality/health evaluation. Techniques such as the multivariable indicator kriging approach (Chapt. 9, this book) should be evaluated for integration of *soft* descriptive and *hard* analytical soil quality/health data into quantitatively based soil quality/health management tools. The Wisconsin correlative report card is currently being applied to experimental sites of farmer-ranked soil quality/health (Chapt. 10, this book). Year-integrated descriptive diagnostic properties used by the farmer were obtained using the site-specific questionnaire or by a site-specific form of the interview guide. Descriptive and analytical properties are transcribed into a standardized form for correlation evaluation.

Management Tools

Management tools for soil quality/health assessment will evolve from results generated by application of the correlative report card to diverse experimental sites.

Predictive Report Card. Establishment of correlative relationships will provide the basis for establishment of predictive groundrules for translation of soil descriptive and analytical properties into output results for the performance of the soil, crop, animal–human, water, and air target systems. The report card at the predictive stage will ideally be based solely on soil properties, and will require an interpretive manual for soil property translation into predicted output performance.

Nontechnical Manual. The descriptive component of the correlative report card should provide the basis of a nontechnical manual that farmers can use to assess and monitor soil quality/health on a continuing basis, to check for signs of deteriorating or improving soil quality/health under normal management, and track potential changes in soil quality/health due to a shift in management practice. As for a patient–doctor partnership, the farmer can keep track of the health of the field, communicate with the scientist in a commonly understood and accepted language, act on scientist recommendations for quantitative analytical assistance and consequent management strategy, and be in an informed position to track the effect of management on soil quality/health. It is likely that the nontechnical soil quality/health manual will include descriptive properties of the soil, crop, animal–human, water, and air target systems, monitored on a continuing basis throughout the year, and will involve guidelines for interpreting interrelationships between descriptive properties of the soil and nonsoil target systems.

SUMMARY

The terms *soil quality* and *soil health* are currently used interchangeably in the scientific literature and popular press. Scientists usually prefer soil quality and farmers usually prefer soil health. Farmer concerns about soil health, a continuing driving force behind soil quality/health research, are rooted in observational field experiences, which translate into descriptive properties of soil and systems supported by or interfacing with soil. Farmer-derived descriptive properties differentiating healthy from unhealthy soils are valuable for (i) defining soil quality/health in meaningful terms, (ii) providing a descriptive component of soil quality/health, and (iii) providing a foundation for developing and validating an analytical component of soil quality/health based on quantifiable chemical, physical, and biological properties, that can be used as a basis for management policy implementation. A soil quality/health interpretive framework is used as a conceptual tool to (i) identify soil and nonsoil (plant, animal–human, water, and air) target systems, (ii) systematize diagnostic properties into major categories within descriptive and analytical groups, (iii) recognize correlation and prediction interrelations between and within target systems, and (iv) establish the functional basis of a definition of soil quality/health. A soil quality/health interview guide and related questionnaires and correlative report card are introduced as information-gathering and analysis tools. These tools not only categorize diagnostic properties characterizing high-quality/healthy and low-quality/unhealthy soils, but also address relationships between and among soft descriptive and hard analytical properties of soil and nonsoil target systems.

ACKNOWLEDGMENTS

Conceptual contributions from Pamela Porter and other members of the Wisconsin Soil Quality/Health Steering Committee are gratefully acknowledged.

The Wisconsin work was supported by the University of Wisconsin (UW) Center for Integrated Agricultural Systems and the Nutrient and Pest Management Program, the UW Agricultural Technology and Family Farm Institute, the Wisconsin Integrated Cropping Systems Trial and the Kellogg Foundation, Wisconsin Department of Agriculture, Trade, and Consumer Protection (WDATCP) Sustainable Agriculture Demonstration Project, the WDATCP Fertilizer Research Council, and the Wisconsin Liming Materials Council. The Washington State work was supported by Northwest Area Foundation, the USDA-STEEP (Solutions To Environmental and Economic Problems) program, and the USDA Dryland Cereal/Legume LISA project.

REFERENCES

Andersen, D. 1992. Farmers, scientists work to develop soil health report card. CIAS Connections 3(2):4–6.

Arshad, M.A., and G.M. Coen. 1992. Characterization of soil quality: Physical and chemical criteria. Am. J. Altern. Agric. 7:25–32.

Beus, C.E., D.F. Bezdicek, J.E. Carlson, D.A. Dillman, D. Granatstein, B.C. Miller, D. Mulla, K. Painter, and D.L. Young (ed.) 1990. Prospects for sustainable agriculture in the Palouse: Farmer experiences and viewpoints. XB 1016. Washington State Univ., Pullman, WA.

Fream, W. 1890. Tilth. p. 95–100. In Soils and their properties. George Bell & Sons, London.

Granatstein, D. 1990. Crop and soil management. p. 5–34. In C.E. Beus et al. (ed.) Prospects for sustainable agriculture in the Palouse: Farmer experiences and viewpoints. XB 1016. Washington State Univ., Pullman, WA.

Haberern, J. 1992a. A soil health index. J. Soil Water Conserv. 47:6.

Haberern, J. 1992b. Coming full circle—the new emphasis on soil quality. Am. J. Altern. Agric. 7:3–4.

Harris, R.F. 1992. Developing a soil health report card. Proc. Wisconsin Fert. Aglime Pest Manage. Conf. 31:245–248.

Harris, R.F., M.J. Garwick, P.A. Porter, and A.V. Kurakov. 1992. Farmer/institution partnership in developing a soil quality/health report card. Participatory On-Farm Research and Education for Agricultural Sustainability. p. 223–224. University of Illinois, Urbana-Champaign, IL.

Heberlein, T.A. 1988. Improving interdisciplinary research: Integrating the social and natural sciences. Soc. Nat. Resour. 1:5–16.

Karlen, D.L., D.C. Erbach, T.T. Kaspar, T.S. Colvin, E.C. Berry, and D.R. Timmons. 1990. Soil tilth: A review of past perceptions and future needs. Soil Sci. Soc. Am. J. 54:153–161.

Larson, W.E., and F.J. Pierce. 1991. Conservation and enhancement of soil quality. In Evaluation for Sustainable Land Management in the Developing World. Vol. 2. ISBRAM Proc. 12 (2). Int. Board for Soil Res. and Management. Bangkok, Thailand.

Papendick, R.I., and J.F. Parr. 1992. Soil quality—the key to a sustainable agriculture. Am. J. Altern. Agric. 7:2–3.

Pawluk, R.R., J.A. Sandor, and J.A. Tabor. 1992. The role of indigenous knowledge in agricultural development. J. Soil Water Conserv. 47:298–302.

Porter, P.A. 1991. Soil biological health: What do Wisconsin farmers think? NPM Field Notes 2:3. Center for Integrated Agric. Systems, Univ. of Wisconsin, Madison, WI.

Porter, P.A., M.J. Garlynd, A.V. Kurakov, and R.F. Harris. 1992. Wisconsin farmer perceptions of soil health. p. 265. *In* Agronomy abstracts. ASA, Madison, WI.

3 The Dynamics of Soil Quality as a Measure of Sustainable Management

W. E. Larson

University of Minnesota
St. Paul, Minnesota

F. J. Pierce

Michigan State University
East Lansing, Michigan

The economic, social, and environmental circumstances of the 1970s and 1980s have compelled American agriculture to assess the sustainability of its land use and production systems. Different perspectives, however, have yielded varying ideas as to what constitutes a sustainable management system. For many, *sustainable management* means stability in production and profitability. For others, the intrinsic goal of sustainable management is to protect and enhance the natural resource base, both biotic and abiotic. For others, maintenance of the social order (e.g., family farm) is essential for sustainability. In short, the concept of sustainability is multidimensional. Thus, sustainable management requires that these and other important concepts must be addressed simultaneously.

Although considerable activity is currently aimed at development and evaluation of sustainable management systems, these efforts are hampered by a lack of agreement on what constitutes credible measures of sustainability. The means to evaluate the sustainability of management systems, both in terms of design and performance, are not yet determined (Larson & Pierce, 1991; Pierce & Larson, 1993). In addition, until measures of sustainability are established, it is not possible to assess how the various components should be weighted in determining sustainability. The latter may deal, for example, with the question of how farm profitability issues should be tempered by environmental concerns and vise versa.

Soil quality is a critical component of sustainable agriculture. While the term *soil quality* is relatively new, it is well known that soils vary in quality and that soil quality changes in response to use and management. The soil system is characterized by attributes that both range within limits and functionally interrelate. Therefore, these attributes can be used to quantify soil quality. Additionally, the soil is an open system, with inputs and outputs, that is bounded (at least artificially) by other systems collectively termed *environment* (Jenny, 1941). Sustainability, then, while multidimensional, is certainly focused both on the quality of the soil resource base and the relationship between its use and management and the environment.

Our hypothesis in this chapter is simple. If it can be agreed that a management system is sustainable only when soil quality is maintained or improved, then a quantitative assessment of the changes in soil quality provides a measure of sustainable management. This chapter defines soil quality and presents an approach to quantify both the inherent and dynamic dimensions of soil quality in terms of minimum data sets (MDS) and pedotransfer functions (PTF) in combination with procedures and models used in *statistical quality control* (SQC) (the introduction of terminology used in statistical quality control has the potential to create some confusion over the use of the term *quality*. However, *quality* is used frequently in SQC literature as a modifier. Therefore, to avoid confusion, references in the text to quality as used in SQC will be italicized). It also explores the concept of designing inherently sustainable land management systems combined with *process quality control procedures* to ensure *quality performance* of the management system design.

APPROACH

Soil Quality and the Sustainable Paradigm

It is important to establish a common paradigm on what constitutes sustainable management. Standards are needed for the design and evaluation of management systems that allow an assessment of their sustainability. We will argue that since the soil system is dynamic, measures of sustainable management should also be dynamic.

An approach frequently employed in the evaluation of sustainable management systems is a *comparative assessment*. A comparative assessment approach is one in which the performance of the system is determined in relation to alternatives. The characteristics and outputs of alternative systems are compared at some time, t, with respect to biotic and abiotic soil system attributes. A decision, based on the difference in magnitude of the measured parameters, is made about the relative sustainability of each system.

Consider an example of a comparative assessment in which two management systems (A and B) occur side by side on the same soil. After some time t, a battery of measurements are taken on the soil and comparisons are made between systems. The differences between systems A and B are presented as evidence of the sustainability of one system relative to the other. An example of a comparative assessment study is given by Reganold (1989) in which he compared two farms near Spokane, WA, managed under conventional and organic farming practices. A brief summary of his comparative data is given in Table 3–1. He concluded from these data that, "in the long-

Table 3-1. Soil physical, chemical, and biological properties of a Naff silt loam soil in the Palouse region of Washington managed under conventional and organic strategy (adapted from Reganold, 1989).

Soil property	Organic farm	Conventional farm
Organic matter (g kg^{-1}) (0–10 cm)	27**	17
Polysaccharides (mg kg^{-1}) 0–10 cm)	1.13*	1.00
CEC (cmol$_c$ kg^{-1}) (0–10 cm)	17.3*	15.6
Total N (mg kg^{-1}) (0–15 cm)	1179*	1066
Ext. P. (mg kg^{-1})	5.5	5.3
Ext. K (cmol$_c$ kg^{-1})	1.2**	0.7
Bulk density (Mg m^{-3})	1.0	1.0
Modulus of rupture (MPa)	0.016*	0.2
Surface horizon thickness (cm)	39.8*	36.7
Depth to Bt horizon (cm)	55.6**	39.8
Color (dry)	10YR 4/2	10YR 5/2, 5/3
Consistency (dry)†	Slightly hard	Hard to very hard
Consistency (wet)†	Friable	Firm
Structure	Granular	Subangular blocky parting to granular

*,** Significant at the 0.05 and 0.01 levels, respectively.
  † Average of summer and fall 1981 and spring 1982 samples. All other samples taken July 1985.

term, the organic farming system was more effective than the conventional farming system in maintaining the tilth and productivity of the Naff soil and in reducing its loss to erosion'' (Reganold, 1989).

On the surface, comparative assessments may appear reasonable, but they are limited for a number of reasons. If only outputs are measured, it provides little information or inference about the process that created the measured condition. It is also not possible to determine if the system that produced the result was poorly designed or operated in a way that was unstable and could not produce the desired output.

In contrast to the comparative assessment approach, we propose a *dynamic assessment* approach, in which the *dynamics* of the system is a measure of its sustainability. In this approach, a management system is assessed in terms of its actual performance determined by measuring attributes of soil quality over time. This does not preclude, however, comparisons of the dynamics of alternative management systems. The dynamic assessment approach is based on principles established in *statistical quality control* (Pierce & Larson, 1993).

Statistical quality control yields a number of important principles that have relevance to the evaluation of the dynamics of soil quality and as a measure of sustainable management (see Pierce & Larson, 1993):

1. *Quality* cannot be "monitored" into the soil.
2. Soil quality is enhanced through *design* of *quality control* to ensure soil quality and identify improvement opportunities to refine these systems.
3. *Process quality control* requires the identification and monitoring of key variables that influence *quality characteristics* of the system.

4. It is important to know the life cycle or pedigree of the process that produces an output to quantify the process variability.
5. Decreasing the variability of the input of a process will tend to decrease the variability of the output (Gilliland, 1990).
6. Statistical tests should be in place so that a decision to adjust the process or not is based on operationally defined standards and not the whims of the people in charge (Gilliland, 1990).
7. As quality is designed into more and more of the process, one can expect the need for quality monitoring of the output will be less.
8. In *statistical quality control*, by controlling the data collection and knowing the time sequence of the data, inferences can be made about the stability of the system over time (Montgomery, 1985).

The dynamic assessment approach should include the following steps:

- Explicit identification of the desired outputs of a management.
- Assessment of the design of the system to determine if it will produce the desired output.
- Identification of the soil quality parameters of importance and establishment of *quality standards*.
- Establishment of the starting point for evaluation of a management system. The condition of the soil at the initiation of the management change is required unless the historical record of the site is good.
- Assessment of the system output to determine if it results from the system design, the system process performance, or both.
- Stabilization of a system process that is *out-of-control*. A stable system of variation is one in which the variation is solely a result of the system in place; there are no special causes of variation (Gilliland, 1990).
- Improvement of the sustainability of a *stable* management system by adjusting it with proper experimental design techniques. This should be attempted only if the system is stable, because tampering with a stable system will make the system less stable (Montgomery, 1985).

What follows is an overview of concepts on soil quality that constitute a framework for its assessment and, subsequently, to achieve sustainable management. We refer the reader to two previous papers on this issue (Larson & Pierce, 1991; Pierce & Larson, 1993).

SOIL QUALITY ASSESSMENT

There are two aspects of the dynamics of soil quality with regard to sustainable management. The first deals with the quantification of soil quality, both in terms of magnitude and dynamics. This aspect deals with how soil quality is changing in response to management. The second deals with design and control of the process by which management systems affect soil quality and, hence, sustainability. The emphasis here is on how the components of a management system and the associated processes acting within a management system are performing with respect to their measured impacts on soil quality.

Quantifying the Dynamics of Soil Quality

The dynamics of soil quality can be quantified by expressing soil quality, Q, as a function of measurable soil atrributes, q_i, measuring the variation of these attributes over time, and evaluating the dynamics of soil quality, dQ/dt, using models or statistical quality control procedures (Larson & Pierce, 1991; Pierce & Larson, 1993).

Larson and Pierce (1991) defined *soil quality* as the capacity of a soil to function, both within its ecosystem boundaries (e.g., soil map unit boundaries) and with the environment external to that ecosystem (particularly relative to air and water quality). Soil quality relates specifically to the ability of soil to function as a medium for plant growth (productivity), in the partitioning and regulation of water flow in the environment, and as an environmental buffer. As a simple operational definition, soil quality means "fitness for use" (Pierce & Larson, 1993). Soil quality, Q, can be defined (Larson & Pierce, 1991) as the state of existence of a soil relative to a standard or in terms of a degree of excellence. It is expressed as a function of attributes of soil quality, q_i, defined as:

$$Q = f(q_{1...n}) \qquad [1]$$

It may be important to think of Q as multidimensional, e.g., as a vector or surface rather than a single point or value. In any event, it is the collective contribution of all q_i values that determine the magnitude of Q.

Although Q is important in land evaluation, sustainable management requires knowledge about changes in soil quality (Pierce & Larson, 1993). The dynamic change in soil quality, dQ/dt, can be defined (Larson & Pierce, 1991) as:

$$\frac{dQ}{dt} = f \left[\frac{\dfrac{(q_{it} - q_{it0})}{q_{it0}} \cdots \dfrac{(q_{nt} - q_{nt0})}{q_{nt0}}}{dt} \right] \qquad [2]$$

An aggrading soil would have a positive dQ/dt and a degrading soil would have a negative dQ/dt. In terms of sustainability, it is our main interest to detect changes in soil quality resulting from a change in land use or management or in measuring the performance of a specific management system in terms of soil quality.

The functional relationship in Eq. [2] is difficult to define, and it is impossible to describe Q in terms of all soil attributes. Therefore, Larson and Pierce (1991) proposed that a MDS, in combination with PTF's, be designed to monitor changes in soil quality. An important aspect of an MDS is that it must include soil attributes in which quantitative attributes can be measured in a short time span to be useful in land use or management decisions. The components of a MDS are selected on the basis of their ease of measurement, reproductibility, and to what extent they represent key variables that control soil quality. It is important to note that any MDS represents a *minimum* set of attributes to be measured to assess soil quality. Other attributes may be part of an extended data set intended for certain investigations. An example of a MDS was described by Larson and Pierce (1991) and a summary is given in Table 3-2. Note that both the type of measurement and a measurement procedure should be standardized, at least within a geographic region. In addition, there are soil parameters that are too costly or difficult to measure that would be desirable in a MDS for soil quality. Fortunately, soil properties are interrelated and can be predicted from other properties using PTF's. A PTF is described by Bouma (1989) as a mathematical function that relates soil characteristics and properties with one another for use in the evaluation of soil quality (Larson & Pierce, 1991). Therefore, PTF's can be used to extend the utility of the MDS to monitor soil quality. Many PTF's occur in the literature and are statistical or empirical in nature. Selected PTF's were discussed by Larson and Pierce (1991) and are given in Table 3-3.

There is no consensus on what a MDS for soil quality should contain. The MDS and PTF's given by Larson and Pierce (1991) represent a starting point. However, a new technical Committee of the International Organization for Standardization (ISO/TC 190) has been recently formed whose scope is "standardization in the field of soil quality, including classification, definition of terms, sampling of soils, and measurement and reporting of soil characteristics" (D.L. Rimmer, 1992, personal communication). A comprehensive effort is also underway by the USEPA Environmental Monitoring and Assessment Program (EMAP) to assess changes in soil quality by monitoring key soil attributes at selected locations nationally (USEPA, 1991).

Table 3-2. Soil attributes and standard methodologies for their measurement to be included as part of a minimum data set (MDS) for monitoring soil quality (adapted from Larson & Pierce, 1991).

Soil attribute	Methodology
Nutrient availability for region	Analytical soil test
Total organic C	Dry or wet combustion
Labile organic C	Digestion with KCl
Texture	Pipette or hydrometer method
Plant-available water capacity	Determined in field best or from water desorption curve
Structure	Bulk density from intact soil cores field measured permeability or Ksat
Strength	Bulk density or penetration resistance
Maximum rooting depth	Crop specific—depth of common roots or standard
pH	Glass electrode–calomel electrode pH meter
Electrical conductivity	Conductivity meter

It is, therefore, essential that working groups be formed in the USA similar to ISO/TC 190,, to develop a MDS and associated PTF's for assessing soil quality.

The soil quality MDS is measured over time to assess the dynamics of soil quality. There are two ways of assessing changes in soil quality:

1. Through the use of computer models to determine how changes in the MDS impact the important functions of soil, such as productivity;
2. Using statistical quality control procedures, a MDS is repeatedly measured over time and the temporal pattern of variation of a MDS parameter or PTF is evaluated (Pierce & Larson, 1993).

The use of models for dynamically assessing soil quality is illustrated by the use of productivity and the loss in productivity with accelerated soil erosion (Pierce & Larson, 1993). Using a modification of a productivity model developed by Kiniry et al. (1983), Pierce et al. (1983) used the PTF's to calculate a normalized sufficiency of pH, bulk density and available water capacity (AWC) for root growth. Once determined, the product of the suffi-

Table 3-3. A limited listing of proposed pedotransfer functions (adapted from Larson & Pierce, 1991).†

PTF no.	Estimate	Relationship
		Chemical
1	Phosphate-sorption capacity	$PSC = 0.4 (AL_{ox} + FE_{ox})$
2	Cation exchange capacity	$CEC = a\ OC + b\ C$
3	Change in organic matter	$\Delta C = a + b\ OR$
		Physical
4	Bulk density	$D_b = b_0 + b_0 + b_1\ OC + b_2\ Si + b_3\ M$
5	Bulk density	Random packing model using particle-size distribution
6	Bulk density	$D_b = f(OC, clay)$
7	Water retention	$q_{10} = b_0 + b_1\ C + b_2\ Sy$
8	Water retention	$q = b_1\ (\%Sa) + b_2\ (\%Si) + b_3\ (\%C) + b_4\ (\%OC)$
9	Random roughness from moldboard plowing	$RR = f(soil\ morphology)$
10	Porosity increase	$P = f(MR, IP, clay, Si, OC)$
		Hydraulic
11	Hydraulic conductivity	$K^S = f(texture)$
12	Seal conductivity	$SC = f(texture)$
13	Saturated hydraulic conductivity	$D_S = f(soil\ morphology)$
		Productivity
14	Soil productivity	$PI = f(D_b, AWHC, pH, E_c, ARE)$
15	Rooting depth	$RD = f(D_b, AWHC, pH)$

† D_{bx} = bulk density; Si = percent silt; M = median sand fraction; OC = organic carbon; C = clay; Sy = $1/D_b$; PSC = phosphate sorption capacity; Al_{ox} = oxalate-extractable Al; Fe_{ox} = oxalate-extractable Fe; OM = organic matter; Sa = sand; MR = moisture ratio; IP = initial porosity; ARE = aeration; AWHC = available water-holding capacity; E_c = electrical conductivity.

ciencies were weighted by a normalized rooting function to calculate a productivity index (PI) as:

$$PI = \sum_{i=1}^{r} (A_i \times C_i \times D_i \times WF)$$ [3]

where A_i is sufficiency of AWC, C_i is the sufficiency of bulk density, D_i is the sufficiency of pH, WF is a weighting factor, and r is the number of horizons in the depth of rooting. Using soil survey data contained in the SOILS-5 data base and land use and erosion data from the National Resources Inventory, the effect of erosion on soil productivity was estimated for the Corn Belt Region of the USA (Pierce et al., 1984). Both the quality (PI) and the change in quality (ΔPI) were estimated using the concepts given in Eq. [1] and [2].

Consider the study reported by Bauer and Black (1992) in which they compared the organic C, bulk density, and AWC of a number of soils from Grant County, North Dakota, that had been cropped to conventionally tilled wheat (*Triticum aestivum* L.), stubble mulch-tilled wheat, grazed grassland, and relict grassland. Twelve nearby sites for each management system were sampled from the same or similar soil type. The soils were then grouped into sandy, medium, and fine texture. The results were summarized by texture and management system. Data from all soils were used in separate regression equations relating bulk density, water concentrations at field capacity, and permanent wilting point to organic C content of the different soil sampling depths. Because the texture varied slightly among the management systems, we used the measured organic C levels to predict bulk density and AWC for the different management systems and textural groupings for the regressions given by the authors. The data for conventionally tilled and relict grazed on a Vebar soil (coarse-loamy, mixed Typic Haploboral) are summarized in Table 3–4.

We calculated PI as an index of soil quality using the bulk density, AWC, and pH. For the sandy soils, PI was 0.71 for conventionally tilled wheat and

Table 3-4. Organic C, bulk density, and available water-holding capacity for sandy-textured North Dakota soils as influenced by two management systems (adapted from Bauer & Black, 1992).

Soil depth	Conventional tilled			Relict grazed		
	OC†	D_b	AWHC	OC	D_b	AWHD
m	g kg^{-1}	Mg m^{-3}	cm^3 cm^{-3}	g kg^{-1}	Mg m^{-3}	cm^3 cm^{-3}
0–0.076	9.9	1.37	0.164	27.9	1.00	0.225
0.076–0.152	9.1	1.39	0.164	17.0	1.23	0.168
0.152–0.305	7.2	1.42	0.162	13.4	1.30	0.168
0.305–0.457	5.6	1.46	0.159	8.9	1.42	0.160
0–0.457	7.4	1.42	0.162	14.9	1.28	0.175

† OC = organic carbon, D_b = bulk density, and AWHC = available water-holding capacity.

0.76 for the relict grassland, a small difference. Except for the surface 0.076 m, the AWC was essentially the same for both treatments. Bulk density was lower in the relict grasslands than the conventionally tilled, but the bulk densities in both treatments were nonlimiting (Pierce et al., 1983). We conclude, as did Bauer and Black (1992), that in these soils the two widely different managements did not greatly affect the productivity of the soil for wheat other than in likely nutrient losses (not measured) in this semi-arid region.

The example from Bauer and Black (1992) and given in Table 3–4 exemplifies the difficulty in determining a standard for Q. Most would argue that the soil from the relict grassland has the higher Q because of the higher organic C and lower bulk density. The reasoning might be that the higher organic C soil would have a higher state of aggregation and, therefore, the soil would be more resistant to wind erosion and rainfall runoff. In this semi-arid climate, runoff is not a major problem. Even though wind erosion is a hazard, it apparently has not materially affected Q (as evidence by PI) after 25 yr of cropping. Water-holding capacity, the most important soil attribute in this semi-arid ecosystem was not influenced significantly, except in the surface 0.076 m layer. Bulk density was higher in the soil with conventionally tilled wheat but was not great enough to limit root growth. Although not measured and not considered in our analysis, the nutrient-supplying power of the soil was probably lower in the soil with conventionally tilled wheat. Thus, degradation of the soil tilled for 25 yr did not lower its performance in a managed wheat ecosystem. Therefore, the degradation appears to have not exceeded *quality standards* for this soil relative to productivity.

One might also argue that the conventionally tilled wheat soil has degraded the atmosphere because the loss in organic C from the soil has supplied CO_2 to the atmosphere and in a small way contributed to the global warming phenomena. Reducing organic C in the 0 to 0.46 m depth from 14.9 to 7.4 g kg^{-1} would supply 143 000 kg ha^{-1} CO_2 to the atmosphere, assuming all C losses resulted from biological oxidation of organic matter.

A second approach to assess the dynamics of soil quality from the MDS is the use of statistical quality control. Pierce and Larson (1993) proposed the use of statistical quality control procedures to describe the dynamics of soil quality and assess sustainable management. The following discussion relies heavily on their paper.

Recall that our interest is mainly in detecting changes in soil quality resulting from a change in land use or management or in measuring the performance of a specific management system in terms of soil quality. Pierce and Larson (1993) stated that statistical tools appropriate for assessing changes in soil quality may be found in the use of control charts. *Control charts* are a standard device used in statistical quality control in the manufactured goods and services industry. The statistical basis for their use is well established and the types and uses of control charts are very diverse (Gilliland, 1990; Montgomery, 1985; Ryan, 1989). Control charts can be thought of as indicators of changing soil quality.

The basic use of control charts is illustrated in Fig. 3–1. Under this procedure, soils would be sampled over time for soil attributes (MDS) represent-

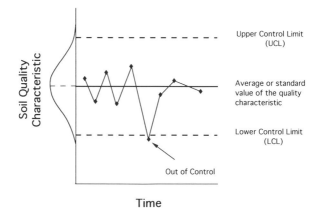

Fig. 3–1. The basic concept of a Shewhart control chart as used for soil quality monitoring (from Pierce & Larson, 1993 after Montgomery, 1985 and Ryan, 1989).

ing *quality parameters* or transformed using PTF's to other *quality parameters.* The upper control limit (UCL) and lower control limit (LCL) are set based on known or desired tolerances, or based on the mean variance obtained from past performance or known through some other means. It may be desirable to design the control limits to represent minimum levels for sustainability, beyond which management cannot be sustained, such as minimum soil organic matter content.

In the simplest case, as long as the sample mean plots within the control limits, the process or system is considered in control. When a sample mean plot is outside the control limits, the process or system is considered out-of-control, i.e., soil quality is changing. Additionally, trends may occur within the control chart, and statistical quality control procedures are available to analyze these trends. Trends may be indicative of instability in the management system or merely characteristic of the process. For example, the data in Fig. 3–2 would be indicative of a system with cyclic variation that operates within the control region (Pierce & Larson, 1993).

The data of Reinert et al. (1991) illustrate the temporal variation in the aggregate stability of a Kalamazoo loam soil (Typic Hapludalf, fine-loamy, mixed, mesic) under varying management: cropped for more than 100 yr and

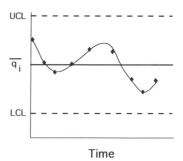

Fig. 3–2. An example of a control chart with variation within the control limits but exhibiting a pattern in the variation of a soil quality parameter (from Pierce & Larson, 1993).

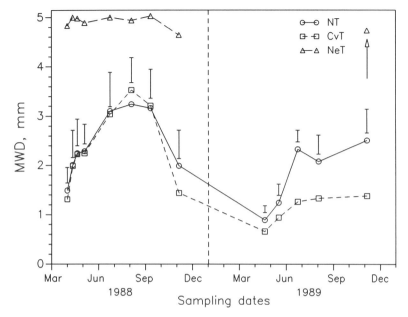

Fig. 3–3. The temporal variation in mean weight diameter (MWD) of soil aggregates sampled from a Kalamazoo loam soil under conventional tillage (CvT), no tillage (NT), and never-tilled (NeT) management for the production of corn (Reinert et al., 1991).

either plowed or no-tilled compared with never tilled, but all recently cropped to corn (*Zea mays* L.) (Fig. 3–3) (Pierce & Larson, 1993). Note that the mean of the mean weight diameter (MWD) is higher and the temporal variation lower when the never-tilled soil is compared with the long cropped systems, regardless of tillage. These data indicate that both the mean and variance of an attribute of soil quality (q), may be affected by a management system and both should be monitored and charted. If the situation in Fig. 3–3 is extrapolated over the long-term (Fig. 3–4), one might expect that no-tillage of the long-term cultivated area would increase the mean and reduce the variance of the MWD of the cultivated soil (Reinert et al., 1991). Conversely, no-till crop production of the never-tilled soil should result in a decrease in the mean and an increase in the variance of MWD over time.

The concept of using control charts for each MDS and PTF parameter is useful in quantifying the dymamics of soil quality. It is likely that, for a given management system, some q_i values may be stable, others *out-of-control*, and others showing trends. A sustainable management system will be characterized by q_i values that are stable over time and, if trends occur in the control charts, they are indicative of an aggrading soil quality, not degrading.

System Design and Process Control

As we indicated earlier, there is a second aspect of the dynamics of soil quality that needs to be addressed with regard to sustainable management.

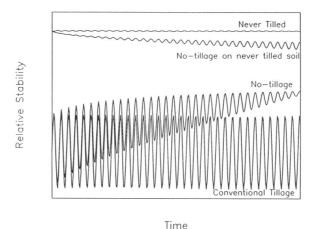

Fig. 3–4. A conceptual illustration of the expected change in relative stability over the long-term of a soil in cropland managed under no-tillage and conventional tillage on previous tilled soil or no-tillage of a never-tilled soil (Reinert et al., 1991).

The first dealt with the quantification of soil quality, both in terms of its magnitude and dynamics. The second deals with the design of management systems, and the process by which management systems affect soil quality, and hence, sustainability.

Pierce and Larson (1993) pointed out that a basic principle in the manufacture of products is that you cannot sample quality into a product. The obvious analogy is that you cannot monitor quality into the soil. Experimental design and process control are required in the manufacture of goods and services to build quality. Therefore, sustainable management requires a deliberate effort to design management systems that are inherently sustainable and the use of process control measures to ensure soil quality and to identify improvement opportunities to refine these design systems. An important outcome is that if quality is built into both the design and the process, quality assurance sampling (monitoring) is minimized (Montgomery, 1985).

An important aspect of process control is that it be in the hands of those managing the process. The manager should be able to interpret the control charts and take appropriate action to adjust the process and bring it back into control. Informing the manager that the outcome of the process is unacceptable is useless in helping the manager achieve the desired quality of the output. Consider the management of crop residues for erosion control as an example of a management practice that relies on monitoring with little regard to the notion of design and process control (Pierce & Larson, 1993). Residue cover is a key factor influencing erosion control. Until recently, interest focused specifically on the amount of residue cover after planting, and it was not uncommon that conservation tillage systems failed to meet the target residue coverage amount of 30% on the erosive landscape positions. The problem is that the standard measure of residue cover occurs after most

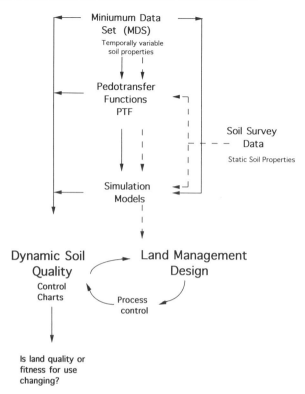

Fig. 3-5. A flow diagram illustrating a procedure for evaluating the sustainability of land management systems (from Pierce & Larson, 1993).

of the management practices that affect residue cover are completed. At this time, it is not possible for the farm manager to alter those practices to ensure the proper residue cover. How would process control measures impact residue cover, if it can be assumed that the design of the system is correct? Because the harvest, tillage, and planting operations impact residue distribution, statistical process controls could be implemented to monitor machinery performance relative to residue management to detect when machinery adjustments are required to achieve desired residue coverage. This level of management is nearly achievable with the site-specific crop management technology currently available (Larson & Robert, 1991). Thus, if the system is designed to meet the intended output, then through process control the output can be reasonably assured.

The overall process of assessing soil quality and evaluating management is illustrated in Fig. 3-5 (Pierce & Larson, 1993). Soil survey data, MDSs, and PTFs provide input to simulation models to design sustainable management systems and establish standards for soil quality. Control charts of various q_i values are monitored with time and used alone or in combination improvement opportunities in the system. Control charts can also be used in combination with MDSs and PTFs to monitor soil quality and serve as thresholds and criteria for *quality standards*.

SUMMARY

Soil quality is defined as the capacity of a soil to function, both within its ecosystem boundaries and with the environment external to that system. Soil quality relates specifically to its ability to function as a medium for plant growth, in the partitioning and regulation of water flow in the environment, and as an environmental buffer.

Soil quality is the state of existence of a soil relative to a standard, or in terms of a degree of excellence. Soil quality is a critical component of sustainable agriculture. A soil management system is sustainable only when it maintains or improves soil quality. However, standards for soil quality and for the design and evaluation of management systems are needed to assess sustainability of management systems. In this chapter, we define soil quality, Q, in terms of soil attributes, q_i's, and propose the use of a MDS and PTFs in conjunction with models and statistical quality control procedures to quantify Q. We favor the use of a dynamic assessment approach to evaluating soil quality in which a management system is assessed in terms of its actual performance rather than in comparison with other systems. The dynamics of soil quality is determined by measuring all parameters of a soil quality MDS over time and assessing their pattern of variation expressed in control charts using SQC procedures. We also recognize the need to apply the SQC concepts of design and process control of inherently sustainable management systems to provide the manager the tools needed to achieve soil quality standards of sustainability.

Two approaches to evaluating sustainable management are (i) comparative and (ii) dynamic. In the comparative approach, two or more management systems are compared and judgements made as to which system is most *sustainable*. Usually, a minimum of attention is given as to the attributes that contribute to the behavior of the systems. In the dynamic approach, which we favor, the system is assessed in terms of its actual performance determined by measuring soil quality parameters over time. The dynamic assessment approach is based on principles established in *statistical quality control*.

The dynamics of soil quality can be identified by expressing soil quality, Q, as a function of measurable soil attributes, q_i values, measuring the variation of these attributes over time, and evaluating the dynamics of soil quality, dQ/dt, using models or *statistical quality control procedures*.

Larson and Pierce (1991) proposed that a MDS, in combination with PTFs, be designed to monitor soil quality changes over time. The MDS must include soil attributes whose changes can be measured in a short period of time. The components of a MDS are selected on the basis of their ease of measurements, reproducibility, and to what extent they represent key variables controlling soil quality. Because soil attributes are interrelated, one attribute can often be predicted from others. Therefore, PTFs can be used to extend the utility of the MDS to monitor soil quality.

Using a monitoring system over time, soil quality changes can be assessed using (i) computer models, or (ii) *statistical quality control* procedures. The latter requires establishment of *quality standards* and *control limits*. The concept of using control charts for each MDF and PTF parameter is attractive. In this way, changes in critical parameters can be identified and corrective action taken to adjust the management system.

REFERENCES

Bauer, A., and A.L. Black. 1992. Organic carbon effects on available water capacity of three soil textural groups. Soil Sci. Soc. Am. J. 56:248-254.

Bouma, J. 1989. Using soil survey data for quantitative land evaluation. Adv. Soil Sci. 9:177-213.

Gilliland, D.C. 1990. Experiences in statistics. Kendall/Hunt Publ. Co., Dubuque, IA.

Jenny, H. 1941. Factors of soil formation. McGraw-Hill Book Co., New York.

Kiniry, L.N., C.L. Scrivner, and M.E. Keener. 1983. A soil productivity index based upon predicted water depletion and root growth. Missouri Agric. Exp. Stn. Res. Bull. 1051.

Larson, W.E., and F.J. Pierce. 1991. Conservation and enhancement of soil quality. *In* Evaluation for sustainable land maangement in the developing world. Vol. 2. IBSRAM Proc. 12(2). Int. Board for Soil Res. and Management, Bangkok, Thailand.

Larson, W.E., and P.C. Robert. 1991. Farming by soil. p. 103–112. *In* R. Lal and F.J. Pierce (ed.) Soil management for sustainability. Soil and Water Conserv. Soc., Ankeny, IA.

Montgomery, D.C. 1985. Introduction to statistical quality control. John Wiley & Sons, New York.

Pierce, F.J., and W.E. Larson. 1993. Developing criteria to evaluate sustainable land management. p. 7–14. *In* J.M. Kimble (ed.) Proc. of the 8th Int. Soil Management Workshop: Utilization of Soil Survey Information for Sustainable Land Use . May 1993. USDA-SCS, National Soil Survey, Lincoln, NE.

Pierce, F.J., W.E. Larson, R.H. Dowdy, and W.A.P. Graham. 1983. Productivity of soils: Assessing long-term changes due to erosion. J. Soil Water Conserv. 38:39–44.

Pierce, F.J., W.E. Larson, R.H. Dowdy, and W.A.P. Graham. 1984. Soil productivity in the Corn Belt: An assessment of erosion's long-term effects. J. Soil Water Conserv. 39:131–136.

Reganold, J.P. 1989. Comparison of soil properties as influenced by organic and conventional farming systems. Am. J. Altern. Agric. 3:144–154.

Reinert, D.J., F.J. Pierce, and M.C. Fortin. 1991. Temporal variation in structural stability induced by tillage. p. 63–72. *In* J.A. Stone et al. (ed.) Soil structure research in eastern Canada. Proc. of the Eastern Canada Soil Structure Workshop, Guelph, Ontario. 10–11 Sept. 1990. Herald Press Limited, Windsor, ON.

Ryan, T.P. 1989. Statistical methods for quality control. John Wiley & Sons, New York.

U.S. Environmental Protection Agency. 1991. An overview of the environmental monitoring and assessment program. EMAP Monitor, January, 1991.

4

A Framework for Evaluating Physical and Chemical Indicators of Soil Quality

Douglas L. Karlen

USDA-ARS
National Soil Tilth Laboratory
Ames, Iowa

Diane E. Stott

USDA-ARS
National Soil Erosion Research Laboratory
West Lafayette, Indiana

Public and private efforts to define and evaluate soil quality are increasing in the USA and throughout the world. The apparent force behind these efforts is public recognition that it is essential to balance the world's finite soil resources with an ever-increasing population, and that soil resources are as vulnerable to degradation as air or water. The rationale is that a quantitative index of soil quality may serve as an indicator of a soil's capacity for sustainable production of crops and animals in an economically sound, socially acceptable, and environmentally friendly manner. These activities are increasing public awareness that soil quality is affected by natural and human-induced processes (Karlen et al., 1992). However, in many areas, inferior soil management practices continue to decrease soil quality through erosion and to create severe off-site damages through sediment, nutrient, and pesticide transport and deposition.

Efforts to define and quantify soil quality are not new, but establishing a consensus with regard to a set of standard conditions to be used for evaluation remains difficult. Unlike water or air quality standards that have been established by legislation using potential human health impact as the primary criterion, soil quality depends on the soil's primary function, which is much more site- and soil-specific. Papendick and Parr (1992) stated that if properly characterized, soil quality should serve as an indicator of the soil's capacity to produce safe and nutritious food, to enhance human and animal health, and to overcome degrative processes. Larson and Pierce (1991) presented a functional definition of soil quality, stating that it is the capacity of a soil to function within ecosystem boundaries and to interact positively with the environment external to that ecosystem. They suggested that the combined physical, chemical, and biological properties of a soil enable it to perform three functions, namely to (i) provide a medium for plant growth, (ii) regulate and partition water flow through the environment, and (iii) serve as an enviornmental filter. They stated that soil quality describes how effectively soils:

1. Accept, hold, and release nutrients and other chemical constituents
2. Accept, hold, and release water to plants, streams, and groundwater
3. Promote and sustain root growth
4. Maintain suitable soil biotic habitat
5. Respond to management and resist degradation.

These proposed functions and characteristics of soil quality are very similar to those proposed by Yoder (1937) as the soil structural characteristics associated with ideal soil tilth. Yoder (1937) defined soil tilth as a blanket term that described the degree of fitness of a soil to support growth and development of plants and concluded that soil structure was a critical component of soil tilth (Karlen et al., 1990). Yoder (1937) suggested that soil structure characteristics essential for ideal tilth would include (i) offering minimum resistance to root penetration, (ii) permitting free intake and moderate retention of rainfall, (iii) providing an optimum soil–air supply with moderate gaseous exchange between soil and atmosphere, (iv) holding to a minimum the competition between air and water for occupancy of pore space volume, (v) providing maximum resistance to erosion, (vi) facilitating placement of green manures and organic residues, (vii) promoting microbiological activity, and (viii) providing stable traction for farm implements.

Soil quality and soil productivity are frequently used synonymously when assessing the long-term impact of soil erosion on crop productivity (Follett & Stewart, 1985; Larson et al., 1990; Verity & Anderson, 1990). This probably occurs because many of the biological, chemical, and physical properties affected by erosion are those that impart a soil's inherent or derived quality (Karlen et al., 1992). However, it may also reflect an anthropocentric attitude that soil is an entity to be used or even mined, rather than a natural resource to be carefully managed.

Soil quality assessments have not been restricted to traditional agronomic crops. Studies by Conry and Clinch (1989) related soil quality to forest species on the Slieve Bloom mountains and foothills in the Central Plain of Ireland. They cited several other reports documenting the importance of soil quality on tree growth and dominant species in many areas. McBride et al. (1990), evaluated potential soil quality impact on forest species on the western end of Manitoulin Island, in Ontario, Canada. For their assessment, they used a portable, noncontacting electromagnetic induction meter to measure apparent electrical conductivity of nonsaline, medium- and coarse-textured forest soils. This information was correlated with tree species and growth as an indication of soil quality.

Soil quality monitoring and evaluation may also be important for addressing environmental problems. For example, Cooper et al. (1985) evaluated environmental issues associated with the U.S. food industry and recommended the development of methods for converting food processing wastes to energy and products to improve soil quality. McBride et al. (1989) reported on soil quality impacts when developing practices for renovation of landfill leachate.

Soil quality also influences the consumer, as shown by Salunkhe and Desai (1988), who reported that soil quality affected nutrient composition of vegetables. Hornick (1992) also identified several soil and management factors that affect soil quality and the nutritional quality of crops.

SOIL QUALITY INDICATORS

Defining soil quality and reaching a consensus with regard to the specific criteria required for its evaluation have been difficult. Presumably this reflects the diversity in properties and characteristics that must be considered because of the numerous alternative uses for the soil resource, including production of agronomic, vegetable, and forestry crops; disposal of agricultural, municipal, and food-processing wastes; or as a filter to protect groundwater resources. To encourage soil quality evaluations, Larson and Pierce (1991) proposed developing a minimum data set (MDS) and suggested using pedotransfer functions (Bouma, 1989) to estimate those parameters if actual measurements were not available. These site-specific, soil quality evaluations could be used to measure changes caused by soil and crop management practices used to control soil erosion or prevent surface and groundwater contamination. They proposed that soil texture, structure, and strength; plant-available water capacity (PAWC); and maximum rooting depth be used as physical indicators of soil quality. They suggested nutrient availability, total organic C, labile organic C, pH, and electrolytic conductivity as chemical indicators.

Larson and Pierce (1991) suggested that all indicator (q_i) measurements could be combined to produce an overall measure of soil quality (Q), a change in quality (dQ) or a change in quality over time (dQ/dt) in response to alternate management practices. They did not state whether all q_i values should be equally weighted or how they might differ for various soil functions or problems.

Soil Conservation Service (SCS) personnel (C.H. Lander, 1992, personal communication) have established several goals related to soil quality. These include (i) identifying parameters that are measurable with current technology, (ii) establishing criteria or values for quantifying those parameters, (iii) developing a structure for evaluating soil quality for both short- and long-term periods, (iv) identifying all management components and their effects on soil quality, and (v) evaluating existing knowledge and research data to determine appropriate indicators and procedures for combining them. The SCS personnel also proposed several physical indicators of soil quality including infiltration, soil texture, aggregation, soil structure, bulk density, plant root development, drainage, permeability, water retention, aeration, available water capacity, capillary water capacity, heat transfer, crusting, tilth, depth to restrictive layers, surface roughness, and soil depth. As chemical indicators, they proposed using cation exchange capacity, fertility, and organic matter content.

At an international conference on the assessment and monitoring of soil quality (Papendick, 1991), infiltration, available water holding capacity, and soil depth were proposed as first-order soil physical properties affecting soil quality, with water stable aggregates, dispersible clay, and bulk density identified as second-order properties. Chemical indicators included pH, salinity, cation exchange capacity, organic matter, and site-specific toxicities such as heavy metals, toxic organics, nitrate, or radioactivity. Conference participants recommended that standard characterization for soil quality include assessments of climate, landscape, soil chemical, and soil physical properties (pH, salinity, organic matter, CEC, texture, bulk density, and available water holding capacity). Several biological indicators were also discussed and are currently being evaluated, but currently none have been agreed on and recommended for routine measurement.

Many groups have suggested similar physical and chemical indicators for assessing soil quality. This includes the Cooperative State Research Service (CSRS), North Central Regional Committee 59, which is focusing on soil organic matter and biological indices of soil quality; a four-state, Northwest Area Foundation project, directed toward development and application of soil quality indices to assess effects of the Conservation Reserve Program on the soil resource; a multi-organization soil quality task force; and a sustainable agriculture and natural resource management (SANREM) consortium, which is developing a research program that will use a *Landscape Approach to Sustainability in the Tropics* (LAST). These and probably several other soil quality activities demonstrate not only interest, but also the difficulty in precisely defining soil quality. For example, an assessment of soil quality for crop production may include the same physical and chemical indicators as an assessment of soil quality relative to soil erosion, but the relative importance of each indicator may be quite different. The critical component that appears absent from many of these investigations is a framework for combining various physical, chemical, and eventually biological properties into either general or specific soil quality indices. Within the framework, it is essential that flexibility be provided to accommodate differences in landscape factors including the drainage or erosion sequence, slope, proximity to an aquifer or other water body, and what the potential land use options may be.

OBJECTIVES

Our primary objective is to propose a framework for evaluating soil quality. The approach, which has also been used to develop a prototype Decision Support System (DSS) to evaluate environmental and economic consequences of alternate farming practices (Yakowitz et al., 1992b), is based on well developed concepts of systems engineering (Wymore, 1993) that have been applied to develop multi-objective decision–support approaches for other natural resource problems. The approach is demonstrated by using a very simple example related to soil erosion by water, because more complex relationships among factors contributing to a site-specific soil quality index remain under development. The simple example illustrates the mechanics of the proposed framework, although the specific weighting functions could actually be determined quantitatively from the sensitivity analyses of the Water Erosion Prediction Project (WEPP) model. We anticipate that this approach will be more useful when tailored to site-specific situations and used to quantify soil quality changes when factors such as short-term economics vs. long-term sustainability, or water erosion vs. deep percolation of chemicals are considered.

For our example, a high soil quality soil with regard to resisting erosion by water must (i) accommodate water entry, (ii) facilitate water transfer and absorption, (iii) resist physical degradation, and (iv) sustain plant growth. Possible soil quality indicators are identified at several levels within the framework for each of these functions. Each indicator is assigned a priority or weight that reflects its relative importance using a multi-objective approach based on principles of systems engineering. Individual scores, which range from 0 to 1, and weights are multiplied and products are summed to provide an overall soil quality rating based on several physical and chemical indicators. The framework and procedure are demonstrated using information collected from an alternative and conventional farm in central Iowa. The use of an expanded framework for assessing effects of other processes, management practices, or policy issues on soil quality is also discussed.

SOIL EROSION PREDICTION TECHNOLOGY

The Water Erosion Prediction Project, which was initiated in 1985, significantly changed soil erosion prediction technology that is available for use in soil and water conservation planning and assessment (Foster & Lane, 1987). The computer model for WEPP is based on fundamentals of infiltration, surface runoff, plant growth, residue decomposition, hydraulics, tillage, management, soil consolidation, and erosoin mechanics (Nearing et al., 1989, 1990a). Several of these factors have been suggested as possible physical indicators of soil quality.

The hillslope version of the WEPP model has capabilities for estimating spatial and temporal distributions of net soil loss and is designed to accommodate spatial variability in topography, surface roughness, soil properties, hydrology, and land use conditions on hillslopes. It can also predict the amount and composition of sediments that leave the land and how those sediments might affect the quality of surface waters. A sensitivity analysis (Nearing et al., 1990b) showed that the 1989 release of WEPP responded primarily to precipitation, rill erodibility, rill residue cover, and rill hydraulic friction forces. Saturated hydraulic conductivity and interrill erodibility were moderately sensitive parameters, whereas soil bulk density, antecedent moisture, sediment characteristics, and several other parameters had less influence on simulated erosion. Later releases of WEPP have shown that interrill cover is also an important factor. The WEPP technology represents the current best understanding of soil erosion by water; therefore, this simulation model could be used to provide input for assessing soil quality as Yakowitz et al. (1992b) have done for water quality modeling, but this was not done for our simple example.

SOIL QUALITY AS RELATED TO SOIL EROSION

Our criteria for a high-quality soil in this simple example were based on a sensitivity analysis of the 1989 WEPP model (Nearing et al., 1990b). They suggested that the most critical functions of a soil with respect to water erosion are (i) to allow water to enter the soil surface, (ii) to facilitate internal transport and absorption of the water, (iii) to resist physical and chemical degradation forces associated with raindrop impact, and (iv) to serve as a medium for plant growth. This final function was critical, not only for the system to be economically viable, but also because plant residues provide the predominant source of rill and interrill cover, which was found to be very important for control of water erosion (Nearing et al., 1990b). Based on this assessment, infiltration, hydraulic conductivity, shear strength, and aggregate stability are among the most critical physical indicators of soil quality with regard to soil erosion. Physical properties such as bulk density, and chemical characteristics including nutrient availability, acidity, electrical conductivity, and salinity can be considered important with regard to sustaining plant growth. Organic matter concentrations and mineralogy are important with regard to aggregate formation and stability, but for our simplified example these will be considered to be of lesser importance.

The extent to which a soil erodes depends on the ability of soil structural units, such as aggregates, to withstand the forces of raindrop impact and surface flow (Meyer, 1981). Several soil properties are known to effect a soil's susceptibility to erosion (Wischmeier & Mannering, 1969). Properties contributing significantly to soil loss variance are soil texture and organic matter content. Measurements of aggregate stability, parent material, and residual effects of crops and management practices are factors correlated to variations in soil loss. Soil properties shown to specifically affect soil detachment include organic matter content (Young & Onstad, 1978), aggregate stability (Young & Onstad, 1978; Luk, 1979; Meyer & Harmon, 1984), and infiltration rates (Poesen & Savat, 1981). Soil shear strength has been used to predict soil erodibility or other soil-related processes (Foster, 1982). A field study by Watson and Laflen (1986) showed a close relation between interrill erosion and soil shear strength after rainfall. Current efforts to develop erosion prediction models that are more processed based than the older predictive technologies include measurements of soil strength indices (Foster & Lane, 1987; Elliot et al., 1989).

A soil aggregate is the basic component of soil structure, and the aggregation process plays a key role in determining the total structure of the soil. Aggregates are formed by physical, chemical, and biological processes, but are stabilized primarily by organic materials. Tisdall and Oades (1982) divided soil organic matter into three major parts, based on their capacity to bind soil particles together into aggregates. The first component was the persistent binding agents, consisting of humic substances (humic and fulvic acids, humins) associated with amorphous Fe and Al and with aluminosilicates, forming organo–mineral complexes in the soil. These agents are associated with microaggregation (aggregates $< 20 \mu m$). From 52 to 98% of the total soil organic matter may be involved in these complexes (Edwards & Bremner, 1967; Hamblin, 1977). The second group consisted of temporary binding agents, such as roots and fungal hyphae, which are primarily associated with macroaggregation (aggregates $> 250 \mu m$). These agents may persist for years, but their presence is highly dependent on soil management and crop rotations. The third component includes the transient binding agents made of plant and microbial polysaccharides and gums, as well as the more complex soil polysaccharides. These agents contribute to micro- (20–250 μm) and macro- ($> 250 \mu m$) aggregation. They are effective for only a short time, from weeks to a few months, and are also affected by soil and crop management.

The extent of aggregation within a soil is also the controlling factor for soil porosity. A well aggregated soil maintains a structure that allows a variety of pore spaces (Lynch & Bragg, 1985). It is the coarse pores, those greater than 75 μm in diam., that control infiltration rates, allowing water to drain under the influence of gravity. On the other hand, fine pores between 0.1 and 15 μm control the capacity of a soil to store plant-available water. Pores of all sizes are necessary for root growth and extension. Roots and mycorrhizal fungi will grow through and even expand small pores, but they cannot penetrate compacted soil. In turn, the presence of fungal hyphae and fine roots impact the size and stability of the aggregates.

Much of the cultivated land around the world is susceptible to various degrees of crusting. A crust forms when surface aggregates disintegrate, filling pore spaces with fine particles. The surface seal that results from this aggregate breakdown impedes water infiltration, leading to runoff and erosion. Such seals may also interfere with seedling emergence, leading to poor crop stands.

Surface seals are less than 2 to 3 mm thick, and are characterized by greater density, finer pores, and water transmission properties that are 2 to 3 orders of magnitude lower than the underlying soil. Hence, they are the principle cause for increased water runoff (Oster & Singer, 1984; Agassi et al., 1981) and soil erosion (Miller & Baharuddin, 1986). Soil susceptibility to sealing depends on numerous soil chemical and physical properties.

Surface seal formation in montmorillonitic soils is very sensitive to the type and amount of cations in the soil matrix and soil solution at the surface. The higher the exchangeable sodium percentage (ESP) and the lower the salt concentration in irrigation waters, the more quickly formed and less permeable the seal (Agassi et al., 1981; Kazman et al., 1983). Soil texture, especially clay content, also affects sealing. Bertrand and Sor (1962) found that the higher the clay content the more aggregated the soil surface remained during rain. Ben-Hur et al. (1985) noted that medium-textured soils, with about 20% clay, were the most sensitive to sealing. Clay mineralogy is another important factor determining sealing. Levy and van der Watt (1988) found that South African kaolinitic soils were less susceptible to seal formation than Israeli smectitic soils. However, highly weathered kaolinitic soils from the southeastern USA were also extremely sensitive to sealing (Miller, 1987).

The effect of soil organic matter on the susceptibility of soils to crust formation and runoff production has received little attention. Generally, organic matter is considered to be a cementing agent that should stabilize soil structure and decrease soil susceptibility to crust formation and surface sealing. Several investigations reviewed by Harris et al. (1966) and more recent studies (Tisdall & Oades, 1982; Dong et al., 1983; Haynes & Swift, 1990, etc.) showed that organic matter content is an important factor in binding particles into water-stable aggregates, generally by forming an organo–mineral complex. Reinterpretation of published data led Sullivan (1990) to propose that organic matter can increase air encapsulation within the soil during water uptake that in turn decreases water uptake rates and prevents slaking and aggregate breakdown. Perfect et al. (1990) found that aggregate stability and clay dispersibility were significantly influenced by antecedent water content. Arshad and Schnitzer (1987) compared a soil that exhibited only slight crusting with a soil that showed severe crusting. They noted that the slightly crusted soil contained more organic matter, and that the organic matter was richer in carbohydrates and proteinaceous material than the severely crusted soil. In addition, the humic acid isolated from the slightly crusted soil was more aliphatic, had lower aromaticity, and contained fewer COOH groups than that of the severely crusted soil.

Organic residues, including plant residues and dead roots, also impact aggregate stability. Plant residues are the major food source for the soil microbial population. Microbial polysaccharides and fungal hyphae are produced during the decomposition process. These cementing and binding materials are extremely important in enhancing soil aggregate stability (Chapt. 16, this book; Molope, 1987).

Aggregation, surface sealing, and porosity are three indicators of soil structure that can have major impact on the critical functions associated with soil quality with regard to soil erosion by water. We have therefore chosen these factors as the primary ones for demonstrating how a conceptual model for evaluating soil quality can be developed. We recognize that this application is relatively simple and that the weighting functions could actually be determined quantitatively from the sensitivity analyses of the WEPP model. However, our goal is to illustate the process being used rather than to deliver a final soil quality assessment procedure. Important long-range goals for this process include the ability to tailor soil quality assessments to site-specific situations or problems and to quantify soil quality impacts if factors such as short-term economics vs. long-term sustainability, or water erosion vs. deep percolation of chemicals are also considered.

CONCEPTUAL SOIL QUALITY MODEL

To develop a quantitative soil quality index as related to soil erosion by water, subjective, qualitative, and quantitative measurements of all appropriate and meaningful chemical and physical indicators must be combined in a consistent and reproducible manner. One possible combination is presented in Table 4-1. Other combinations are undoubtedly possible, and the specific physical and chemical indicators chosen to evaluate a specific function may differ depending on the physiographic region (climate, topography, etc.) for which a soil quality assessment is being made, and the type of water erosion (rill, interrill, gully, etc.) that dominates the landscape. Choices of specific indicators may also depend on information available, or on the confidence in various pedotransfer functions, if a minimum data set approach (Larson & Pierce, 1991) is utilized.

The functions chosen for our conceptual model of soil quality as related to erosion by water were derived primarily from the sensitivity analysis of WEPP 89 (Nearing et al., 1990b); however, suggestions from various soil quality working groups have also been included. A systems engineering technique was applied to define a soil quality rating with regard to erosion by water to provide a mechanism for assigning relative weights to each function (Table 4-1). Within each level, relative weights will be given to each indicator. These weights may change over time or location, depending on priorities or uncontrollable factors, but the approach or framework for developing a quantitative procedure for evaluating soil quality is constant.

We suggest that the primary function of a soil with high quality, relative to erosion by water, is to accommodate entry of the water into the soil matrix through the infiltration rate and capacity. If the water can enter the soil, it will not run off, and thus initiate the erosion process. Based on this rationale, we suggest that this function be given a weight of 0.50 or 50% (Table 4-1). For water to be able to enter the soil matrix, resistance of the surface structure to degradation and transport away from the surface are assumed to be the next two most critical functions. We have thus assigned these two functions weights of 0.35 and 0.10, or 35 and 10%, respectively. The remaining 5% is assigned to the function of sustaining plant growth. This function interacts with resisting degradation at the surface, since rill residue cover is a sensitive parameter (Nearing et al., 1990b). Residue cover also affects entry and transport functions, since plants may provide macropores, racks, and other channels for water entry.

For this simplified example, the ability of the soil to sustain plant growth is assumed to be a less important than the processes contributing to water entry and transport or to aggregate formation and stability. Obviously, these assumptions and weights would not be true if soil quality were being assessed with regard to crop productivity. However, the proposed framework can be

Table 4–1. Soil quality functions and indicators related to erosion by water.

				Indicator			
Function	Weight	Level I	Weight	Level II	Weight	Level III	Weight
Accommodate water entry	0.50	Infiltration rate	1.0	Surface crust	0.20	Texture	0.50
						Thickness	0.20
						Strength	0.20
						Formation rate	0.10
				Surface roughness	0.20		
				Crop residue cover	0.50		
				Macropores	0.10	Plant roots	0.40
						Earthworms	0.60
Facilitate water transfer and absorption	0.10	Hydraulic conductivity	0.60	Soil texture	0.50		
				Capillary water content	0.30		
				Bulk density	0.20		
		Porosity	0.15				
		Macropores	0.25	Plant roots	0.40		
				Earthworms	0.60		
Resist degradation	0.35	Aggregate stability	0.80	Mineralogy	0.20	Physical bonding	0.50
						Chemical bonding	0.50
				Soil carbohydrates	0.35	Labile organic C	0.60
						Recalcitrant organic C	0.40
				Microbial biomass	0.30	Fungal hyphae	0.50
						Actinomycetes	0.30
						Bacteria	0.20

Function	wt	Process	wt	Indicator	wt		wt
Sustain plant growth	0.05	Shear strength	0.10	Available cations	0.15		
		Soil texture	0.05				
		Heat transfer capacity	0.05				
		Rooting depth	0.25	Restrictive layer depth	0.70		
				Texture	0.30		
		Water relations	0.35	PAWC†	0.70		
				Drainage	0.20		
				Organic C	0.10		
		Nutrient relations	0.30	pH	0.15		
				Organic C	0.25		
				Macronutrients	0.40	N	0.50
						P	0.25
						K	0.25
				Ca, Mg, S	0.10		
				B, Cu, Fe, Mn, Zn	0.10		
		Chemical barriers	0.10	Salinity	0.50		
				Heavy metals	0.15		
				Organics‡	0.25		
				Radioactive	0.10		

† PAWC = plant-available water capacity.
‡ Includes contamination by pesticides.

easily modified and used to compute a series of soil quality indices relative to various problems.

After assigning relative weights to the functions necessary for a soil to resist erosion by water, physical and chemical indicators useful for evaluating those functions can be identified and prioritized. To quantify soil quality relative to the function of accommodating water entry into a soil, we suggest the best indicator would be a direct measure of the infiltration rate. However, this indicator is costly and time-consuming to measure; therefore, it may be more feasible to evaluate how well a soil functions to accommodate water entry with a series of second-level physical indicators (Table 4–1). We suggest surface crusting, surface roughness, crop residue cover, and the presence of macropores as possible second-level indicators for assessing this function. Once again, relative weights for each of these second-level indicators must be determined and assigned. The process is repeated to provide a third level of indicators, and if desirable it can be continued further.

With regard to facilitating water transfer and absorption, soil structure, hydraulic conductivity, depth to restrictive layers, presence of macropores, and internal drainage characteristics are the primary indicators that could be used to assess the quality of a soil. Following a pattern similar to that demonstrated by Larson and Pierce (1991), these Level 1 indicators have been assigned several Level 2 measurements for use as indicators at that level.

A third function of a soil with high quality relative to resisting erosion by water is its ability to resist structural degradation. Critical indicators for assessing this function include measurements of aggregate stability, soil texture, shear strength, and heat transfer capacity. Second-level indicators relative to this soil quality function is where critical chemical and biochemical indicators are first seen in this conceptual model (Table 4–1).

Our fourth function, the ability to sustain plant growth, is much more dependent on several of the chemical indicators than the other functions. However, this primarily reflects the soil quality assessment problem that was chosen. If this assessment had been made relative to crop production, ground-water quality, or even food safety, indicators identified in this section (Table 4–1) would have been much different with very different weights.

EVALUATION MECHANICS

Having identified critical soil functions and potential physical and chemical indicators that could be used to assess soil quality relative to its ability to resist erosion by water, it is essential to develop a mechanism to combine the distinctly different functions and indicators. This can be done by using standard scoring functions (SSF) (Fig. 4–1) that were developed for systems engineering problems (Wymore, 1993). A similar approach has been used for multi-objective problem solving by Yakowitz et al. (1992a, b).

Use of standard scoring functions enables the user to convert numerical or subjective ratings to unitless values on a 0 to 1 scale. The procedure begins by selecting the appropriate physical and chemical indicators of soil qual-

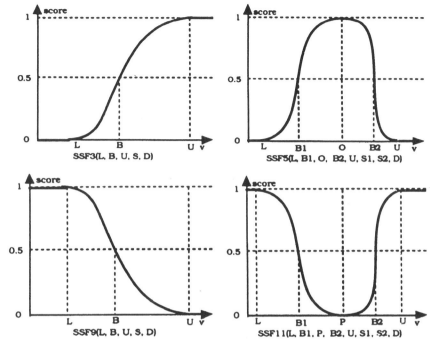

Fig. 4-1. General shapes for standard scoring functions. The upper left indicates "more is better," the upper right "an optimum range," the lower left "less is better," and the lower right "an undesirable range." The letters *L, B, U, and S* refer to the lower threshold, baseline, upper threshold, and slope values, respectively. The *D* value would be the domain over which the function is described (adapted from Wymore, 1993).

ity that affect a particular function related to soil erosion. An appropriate scoring function (Fig. 4-1) and realistic baseline and threshold values for each indicator must be established. Information from experts, specific data bases, or general knowledge can be used for these processes, but for future reference or modification, the source and reason for selection must be documented. For example, Lane et al. (1991) and Yakowitz et al. (1992a, b) used computer simulation models to obtain values for criteria that were important to their problems.

All indicators affecting a particular function are grouped together as shown in Table 4-1 and assigned a relative weight based on importance. All weights within each level must sum to 1.0 or the decimal equivalent of 100%. Four of the most common shapes for scoring functions are presented in Fig. 4-1 and are often referred to as *more is better, less is better, an optimum range,* and *an undesirable range* (Wymore, 1993). After scoring each factor, the value is multiplied by the appropriate weight. When all indicators for a particular function have been scored, the user has a matrix that can be summed to provide a soil quality rating as related to erosion by water.

SOIL QUALITY ASSESSMENT FOR TWO IOWA FARMS

To evaluate our approach for assessing soil quality as related to soil erosion by water, recent results were taken from an on-going comparison of conventional and alternative farming practices in central Iowa. The two farms, located near Boone, IA, are owned by Mr. and Mrs. Richard Thompson and Mr. and Mrs. Eugene Baker. The Thompson farm was discussed in depth as Case Study V in the National Research Council's (NRC) report on Alternative Agriculture (NRC, 1989), but that report was criticized because, in general, it lacked quantitative data for the various case studies (CAST, 1990). Since that time, we have evaluated profile N concentrations, earthworm populations, infiltration, and aggregate stability on both farms (Karlen & Colvin, 1992; Berry & Karlen, 1993; Jordahl & Karlen, 1993; Logsdon et al., 1993).

With regard to evaluating soil quality within these adjacent fields, our measurements showed that the soil map units were similar and that they have similar geomorphic, hydrologic, and sedimentologic relationships (Steinward, 1992). However, soil aggregates collected from fields on the Thompson farm were always more stable in water than those from the Baker farm (Jordahl, 1991).

For the function of resisting degradation, we will assume that a *more is better* relationship (Fig. 4–1) is appropriate for assessing effects of soil aggregation. Therefore, since field-moist samples from the conventional (Baker) farm had wet aggregate stability ratings of approximately 50% (Jordahl & Karlen, 1993), we will assign that value as a baseline or 0.5 rating. Water stability of soil aggregates from the Thompson (alternative) farm was approximately 80%, and would have a 0.80 rating. Soil texture for samples being compared was similar, whereas shear strength and heat transfer capacity were not measured; therefore, they will be assumed to be constant for this example. Soil quality with regard to resisting degradation could therefore be rated $0.80 \times 0.8 + 0.2 = 0.84$ for the Thompson farm and $0.50 \times 0.8 + 0.2 = 0.60$ for the Baker farm.

Steady-state infiltration measurements for Clarion loam in 1989 and 1991 (Logsdon et al., 1993) suggested that median rates (m s^{-1}) on the alternative (Thompson) field were 30 to 60% higher than for the conventional (Baker) field. Using a *more is better* relationship (Fig. 4–1) with no difference as the baseline (0.5 value, -100% as the lower threshold, and $+100\%$ as the upper threshold value), the Baker fields would be given a score of 0.5 whereas the Thompson fields would be given an average score of 0.73 for the 2 yr. Soil quality with regard to accommodating water entry could therefore be rated $1.0 \times 0.73 = 0.73$ for the Thompson farm and $1.0 \times 0.5 = 0.5$ for the Baker farm.

Corn yields in 1989 were similar for both farming systems (Karlen & Colvin, 1992) and measurements of hydraulic conductivity, porosity, and macropores have not been made at this site. Assuming these parameters are equivalent for both farms, they will be assigned baseline scores of 0.50.

An overall soil quality rating with regard to soil erosion by water can now be computed by summing the weighted score for each critical function using the procedure outline in Eq. [1].

$$\text{Soil quality } (Q) = q_{we} \ (wt) \times q_{wt} \ (wt) + q_{rd} \ (wt) + q_{spg} \ (wt) \quad [1]$$

where

q_{we} = level 1 rating for accommodating water entry
q_{wt} = level 1 rating for water transport and absorption
q_{rd} = level 1 rating for resisting degradation
q_{spg} = level 1 rating ror supporting plant growth
wt = weighting factor for each function (Table 4–1)

For this hypothetical example, the alternative farm would be rated 0.5 × 0.73 + 0.1 × 0.5 + 0.35 × 0.84 + 0.05 × 0.5 = 0.73, while the conventional farm would be rated 0.5 × 0.5 + 0.1 × 0.5 + 0.35 × 0.6 + 0.05 × 0.5 = 0.54. Undoubtedly, this procedure will require extensive refinement, both in the techniques used to score the various physical and chemical measurements and with regard to weights assigned to each critical function and measurement. However, our purpose was to demonstrate a systems engineering approach that could be used to provide an initial framework for assessing soil quality with regard to the specific problem of soil erosion by water. This approach appears to offer an opportunity to quantify a concept that to date has only been addressed in qualitative and conceptual manner.

SOIL QUALITY EVALUATIONS AT OTHER SCALES

As discussed throughout this chapter, soil quality evaluations will most likely have to be made with regard to specific problems. Our assessment in this report is focused on the problem of soil erosion by water; however, we suggest that a similar framework could be used to develop a soil quality index relative to other problems, such as crop production, wind erosion, groundwater quality, surface water quality, air quality, or food safety. After evaluating soil quality for several problems, a similar framework could be developed to compute soil quality indices for numerous problems at lower and higher scales of evaluation.

Table 4–2 shows how soil quality assessments might be developed to evaluate soil quality impacts at various scales. At the process level, soil quality evaluations might focus on specific questions with regard to aggregate formation and stability; impacts of crop rotation and vertical integration on soil quality might be evaluated at the farm management level; national policy implications on soil quality might affect decisions such as how to structure the long-term management of Conservation Reserve Program (CRP) lands; or at an international scale, relationships could be developed to relate soil quality to socio-economic problems such as expansion of agriculture onto fragile lands in response to increasing population. We propose that it is possible to assess soil quality at the multiple scales shown in Table 4–2, provid-

Table 4-2. Soil quality assessments at various scales of evaluation.

Scale of evaluation	Cause		Consequences		Type of assessment
	Socio-economic	Biophysical	Biophysical	Socio-economic	
Process	Lack of knowledge	Understanding of components	Understanding of soil aggregation	Suboptimum resource management	Process-based soil quality index
Problem	Erosoin by water	Water entry, transmission, soil properties, crop growth	Degraded soil resource base	Increased production cost	Site-specific soil quality index
Management	Farm economics	Crop rotation/vertical integration	Monoculture production/animal waste	Decreased environmental quality	Farm-specific soil quality index
National policy	Urban demand for decreased soil erosion	Soil erosion and sediment deposit	Development of CRP program†	Increased cost to taxpayers	National, policy-based soil quality index
International policy	Increasing human population	Expansion of agriculture onto fragile lands	Increased soil erosion and decreased yield	Inadequate nutrition and increased infant mortality	International, policy-based soil quality index

† CRP = Conservation Reserve Program.

ed a consistent framework and approach are used. Our rationale is that assessments at each level can be viewed as an expansion or combination of factors influencing the assessment at higher or lower levels, respectively. The critical requirement for this approach is to reach consensus on what actual measurements will be valid, reliable, sensitive to change, replicable, and accessible at the various levels of assessment. Finally, through interaction with people who may be affected in any way by a soil quality assessment at a particular level, appropriate weights and priorities can be established.

SUMMARY AND CONCLUSIONS

Quantitative assessments of soil quality may be useful for optimizing land use plans. However, before assessments can be used this way, valid, reliable, sensitive, repeatable, and accessible indicators must be identified and a framework for overall evaluation of soil quality must be developed. Recognizing that several biological indicators are currently being evaluated, but are not yet recommended for routine measurement, this report focuses on physical and chemical indicators of soil quality, as related to soil erosion by water. Our objective is to illustrate a procedure that can be tailored to site-specific situations and used to quantify soil quality impacts, even when tradeoffs between short-term economics vs. long-term sustainability, or water erosion vs. deep percolation of chemicals are considered. For our example, a soil with high quality must (i) accommodate water entry, (ii) facilitate water transfer and absorption, (iii) resist physical degradation, and (iv) sustain plant growth. Potential soil quality indicators that help quantify these functions are identified, assigned a priority or weight that reflects its relative importance, and scored using a systems engineering approach. Individual scores are used to compute an overall soil quality rating based on several physical and chemical indicators. The framework and procedure are demonstrated using data collected from an alternative and conventional farm in central Iowa. For our hypothetical example, a Clarion loam (fine-loamy, mixed, mesic Typic Hapludolls) in the alternative fields would have a soil quality rating of 0.73 compared with 0.54 for the same soil in the conventionally farmed fields when evaluated with regard to soil erosion by water.

Efforts to quantify soil quality are increasing throughout the USA and around the world. Several currently used and emerging physical and chemical indicators have been identified in this chapter. The framework we have suggested may be useful as a model for developing quantitative assessments of soil quality at various scales of evaluation. The procedure for developing the framework is based on the multi-objective analysis principles of systems engineering and focuses on identifying critical functions for a specific problem, process, management practice, or policy issue. Several physical and chemical measurements that can be made at different levels of investigation are identified as a method for quantifying system response with regard to those functions. Each parameter, as well as the functions, are given an appropriate priority or weight and used to compute a soil quality index with regard to a specific problem, process, practice, or policy issue. We conclude that this procedure could be adapted for quantitatively evaluating soil quality for several scales of investigation.

REFERENCES

Agassi, M., I. Shainberg, and J. Morin. 1981. Effect of electrolyte concentration and soil sodicity on infiltration rate and crust formation. Soil Sci. Soc. Am. J. 45:858-851.

Arshad, M.A., and M. Schnitzer. 1987. Characteristics of the organic matter in a slightly and in a severely crusted soil. Z. Pflanzenerndhr. Bodenk. 50:412-416.

Ben-Hur, M., I. Shainberg, D. Bakker, and R. Keren. 1985. Effect of soil texture and $CaCO_3$ content on water infiltration in crusted soil as related to water salinity. Irrig. Sci. 6:281-294.

Berry, E.C., and D.L. Karlen. 1993. Comparison of alternate farming systems: II. Earthworm population density and species diversity. Am. J. Altern. Agric. 8:21-26.

Bertrand, A.R., and K. Sor. 1962. The effects of rainfall intensity on soil structure and migration of colloidal materials in soils. Soil Sci. Soc. Am. Proc. 26:297-300.

Bouma, J. 1989. Using soil survey data for quantitative land evaluation. Adv. Soil Sci. 9:177-213.

Council for Agricultural Science and Technology. 1990. Alternative agriculture scientists' review. Spec. Publ. 16. CAST, Ames, IA.

Conry, M.J., and P. Clinch. 1989. The effect of soil quality on the yield class of a range of forest species grown on the Slieve Bloom Mountain and foothills. Forestry 62:397-407.

Cooper, J.L., K. Dostal, and P.A. Carroad. 1985. Report of the workshop sessions on environmental issues. Food Tech. 39:33R-35R. Inst. of Food Technologists, Chicago.

Dong, A., G. Chester, and G.V. Simsiman. 1983. Soil dispersibility. Soil Sci. 136:208-212.

Edwards, A.P., and J.M. Bremner. 1967. Microaggregates in soils. J. Soil Sci. 18:64-73.

Elliot, W.J., A.M. Liebenow, J.M. Laflen, and K.D. Kohl. 1989. A compendium of soil erodibility data from WEPP cropland soil field erodibility experiments: 1987 & 88. NSERL Rep. 3. Natioal Soil Erosion Res. Lab., USDA-ARS, West Lafayette, IN.

Follett, R.F., and B.A. Stewart (ed.). 1985. Soil erosion and crop productivity. ASA, CSSA, and SSSA, Madison, WI.

Foster, G.R. 1982. Modeling the erosion process. p. 297-360. In C.T. Haan (ed.) Hydrologic modeling of small watersheds. ASAE Monogr. 5. Am. Soc. of Agric. Eng., St. Joseph, MI.

Foster, G.R., and L.J. Lane. 1987. User requirements: USDA water erosoin prediction project (WEPP). NSERL Rep. 1. National Soil Erosion Res. Lab., USDA-ARS, W. Lafayette, IN.

Hamblin, A.P. 1977. Structural features of aggregates in some East Anglian silt soils. J. Soil Sci. 28:23-28.

Harris, R.F., G. Chesters, and O.N. Allen. 1966. Dynamics of soil aggregation. Adv. Agron. 18:107-169.

Haynes, R.J., and R.S. Swift. 1990. Stability of soil aggregates in relation to organic constituents and soil water content. J. Soil Sci. 41:73-83.

Hornick, S.B. 1992. Factors affecting the nutritional quality of crops. Am. J. Altern. Agric. 7:63-68.

Jordahl, J.L., and D.L. Karlen. 1993. Comparison of alternate farming systems: III. Soil aggregate stability. Am. J. Altern. Agric. 8:27-33.

Karlen, D.L., D.C. Erbach, T.C. Kaspar, T.S. Colvin, E.C. Berry, and D.R. Timmons. 1990. Soil tilth: A review of past perceptions and future needs. Soil Sci. Soc. Am. J. 54:153-161.

Karlen, D.L., N.S. Eash, and P.W. Unger. 1992. Soil and crop management effects on soil quality indicators. Am. J. Altern. Agric. 7:48-55.

Karlen, D.L., and T.S. Colvin. 1992. Alternative farming system effects on profile nitrogen concentrations on two Iowa farms. Soil Sci. Soc. Am. J. 56:1249-1256.

Kazman, Z., I. Shainberg, and M. Gal. 1983. Effect of low levels of exchangeable Na and applied phosphogypsum on the infiltration rate of various soils. Soil Sci. 35:184-192.

Lane, L.J., M. Asce, J.C. Ascough, and T.E. Hakonson. 1991. Multiobjective decision theory—decision support systems with embedded simulation models. p. 445-451. In W.F. Ritter (ed.) Proc. of the 1991 National Conf. on Irrigation and Drainage, Honolulu, HI. 22-26 July 1991. Am. Soc. of Civil Eng., New York.

Larson, W.E., G.R. Foster, R.R. Allmaras, and C.M. Smith (ed.). 1990. Proc. Research Issues in Soil Erosion/Productivity, Bloomington, NM. 13–15 Mar. 1989. Univ. of Minnesota, St. Paul, MN.

Larson, W.E., and F.J. Pierce. 1991. Conservation and enhancement of soil quality. *In* Proc. of the Int. Workshop on Evaluation for Sustainable Land Management in the Developing World. Vol. 2. IBSRAM Proc. 12(2). Int. Board for Soil Res. and Management, Bangkok, Thailand.

Levy, G.J., and H.H. van der Watt. 1988. The effects of clay mineralogy and soil sodicity on the infiltration rates of soils. S. Afr. J. Plant Soil 5:92–96.

Logsdon, S.D., J.K. Radke, and D.L. Karlen. 1993. Comparison of alternate farming systems: I. Infiltration techniques. Am. J. Altern. Agric. 8:15–20.

Luk, S.H. 1979. Effect of soil properties on erosoin by wash and splash. Earth Surf. Processes Landforms 4:241–255.

Lynch, J.M., and E. Bragg. 1985. Microorganisms and soil aggregate stability. Adv. Soil Sci. 2:133–171.

McBride, R.A., A.M. Gordon, and P.H. Groenevelt. 1989. Treatment of landfill leachate by spray irrigation: An overview of research results from Ontario, Canada: II. Soil quality for leachate disposal. Bull. Environ. Contam. toxicol. 42:518–525.

McBride, R.A., A.M. Gordon, and S.C. Shrive. 1990. Estimating forest soil quality from terrain measurements of apparent electrical conductivity. Soil Sci. Soc. Am. J. 54:290–293.

Meyer, L.D. 1981. How rain intensity affects interrill erosion. Trans. ASAE 24:1472–1475.

Meyer, L.D., and W.C. Harmon. 1984. Susceptibility of agricultural soils to interrill erosion. Soil Sci. Soc. Am. J. 48:1151–1157.

Miller, W.P. 1987. Infiltration and soil loss of three gypsum amended ultisols under simulated rainfall. Soil Sci. Soc. Am. J. 51:1314–1320.

Miller, W.P., and M.K. Baharuddin. 1986. Effect of dispersible clay on infiltration and soil loss of southeastern soils. Soil Sci. 142:235–240.

Molope, M.B. 1987. Soil aggregate stability: The contribution of biological and physical process. S. Af. J. Plant Soil 4:121–126.

National Research Council. 1989. Crop–livestock farming in Iowa: The Thompson farm. p. 308–323. *In* Alternative agriculture. National Academy Press, Washington, DC.

Nearing, M.A., G.R. Foster, L.J. Lane, and S.C. Finkner. 1989. A process-based soil erosion model for USDA-water erosion prediction project technology. Trans. ASAE 32:1587–1593.

Nearing, M.A., L.J. Lane, E.E. Alberts, and J.M. Laflen. 1990a. Prediction technology for soil erosion by water: Status and research needs. Soil Sci. Soc. Am. J. 43:1072–1711.

Nearing, M.A., L. Deer-Ascough, J.M. Laflen. 1990b. Sensitivity analysis of the WEPP hillslope profile erosion model. Trans. ASAE 33:839–849.

Oster, J.D., and M.J. Singer. 1984. Water penetration problems in California soils. Univ. of California, Davis, CA.

Papendick, R.I. (ed.). 1991. International conference on the assessment and monitoring of soil quality: Conference report and abstracts. Rodale Inst., Emmaus, PA.

Papendick, R.I., and J.F. Parr. 1992. Soil quality—the key to a sustainable agriculture. Am. J. Altern. Agric. 7:2–3.

Perfect, E., B.D. Kay, W.K.P. van Loon, R.W. Sheard, and T. Pojasok. 1990. Factors influencing soil structural stability within a growing season. Soil Sci. Soc. Am. J. 54:173–179.

Poesen, J., and J. Savat. 1981. Detachment and transportation of loose sediments by raindrop splash: II. Detachability and transportability measurements. Catena 8:19–41.

Salunkhe, D.K., and B.B. Desai. 1988. Effects of agricultural practices, handling, processing, and storage on vegetables. p. 23–71. *In* E. Karmas and R.S. Harris (ed.) Nutritional evaluation of food processing. Van Nostrand Reinhold, New York.

Steinwand, A.L. 1992. Soil geomorphic, hydrologic, and sedimentologic relationships and evaluation of soil survey data for a Mollisol catena on the Des Moines Lobe, central Iowa. Ph.D. diss. Iowa State Univ., Ames, IA.

Sullivan, L.A. 1990. Soil organic matter, air encapsulation and water stable aggregation. J. Soil Sci. 41:529–534.

Tisdall, J.M., and J.M. Oades. 1982. organic matter and water-stable aggregates in soil. J. Soil Sci. 33:141–163.

Verity, G.E., and D.W. Anderson. 1990. Soil erosion effects on soil quality and yield. Can. J. Soil Sci. 70:471–484.

Watson, D.A., and J.M. Laflen. 1986. Soil strength, slope and rainfall intensity effects on interrill erosion. Trans. ASAE 29:98–102.

Wischmeier, W.H., and J.V. Mannering. 1969. Relation of soil properties to its erodibility. Soil Sci. Soc. Am. Proc. 33:131–137.

Wymore, A.W. 1993. Model-based systems engineering: An introduction to the Mathematical Theory of Discrete Systems and to the Tricotyledon Theory of System Design. CRC Press, Boca Raton, FL.

Yakowitz, D.S., L.J. Lane, J.J. Stone, P. Heilman, R.K. Reddy, and B. Imam. 1992a. Evaluating land management effects on water quality using multi-objective analysis within a decision supporting system. Am. Water Resour. Assoc. 1st Int. Conf. on Ground Water Ecology, Tampa, FL. 26–29 Apr. 1992. AWRA, Bethesda, MD.

Yakowitz, D.S., L.J. Lane, M. Asce, J.J. Stone, P. Heilman, and R.K. Reddy. 1992b. A decision support system for water quality modeling. p. 188–193. *In* Water resources planning and management. Proc. of the Water Resources Sessions/Water Forum '92, Baltimore, MD. 2–6 Aug. 1992. Am. Soc. of Civil Eng., New York.

Yoder, R.E. 1937. The significance of soil structure in relation to the tilth problem. Soil Sci. Soc. Am. Proc. 2:21–33.

Young, R.A., and C.A. Onstad. 1978. Characterization of rill and interrill eroded soil. Trans. ASAE 21:1126–1130.

5   ·Microbial Indicators of Soil Quality[1]

R. F. Turco

Laboratory for Soil Microbiology
Purdue University
West Lafayette, Indiana

A. C. Kennedy

USDA-ARS
Pullman, Washington

M. D. Jawson

Kerr Laboratory
USEPA
Ada, Oklahoma

The quality of soil can impact land use, sustainability, and productivity. Human and animal health is closely linked to soil productivity and environmental quality. Soil quality investigations are needed to provide information for management and regulatory decisions. Soil organisms contribute to the maintenance of soil quality in that they control the decomposition of plant and animal materials, biogeochemical cycling including N_2 fixation, the formation of soil structure, and the fate of organics applied to soil. Soil microorganisms are potentially one of the most sensitive biological markers available and should, therefore, be useful for classification of disturbed or contaminated systems. Soil microbial processes are an integral part of soil quality, and a better understanding of these processes and microbial community structure is needed. Microbial analysis has often been neglected, mainly because of the perceived difficulty of analysis. A high-quality soil is thought of as being biologically active and containing a stable cross-section of microorganisms. However, as Lee (1991) stated, "for soil organisms, in no case has the full diversity of a soil population been determined, or is it known how optimum biodiversity for an agricultural ecosystem might be defined, or if it were defined, how it might be compared with a similarly defined optimum for a non-agricultural ecosystem." As far back as 1916 (Lipman et al., 1916),

---

[1]Paper no. 13,687 of the Purdue Agricultural Experiment Station series.

it was argued that a lower productive capacity in a soil (a lower soil quality) was a reflection of a neglected microbiological machinery within that soil.

Microbial populations apparently can provide advance evidence of subtle changes in soil long before it can be accurately measured by changes in organic matter (Powlson et al., 1987). Increases in specific populations are seen with straw incorporation (Mukherjee & Gaur, 1980). Long-term management practices of agricultural lands also influence microbial activities (Martynuik & Wagner, 1978; Doran, 1980; Bolton et al., 1985; Ramsay et al., 1986). The ecology of root × microbe interactions after minimum tillage practices is different from the ecology seen after moldboard plowing and other seedbed preparations. Changes in soil physical and chemical properties resulting from tillage greatly alter the soil matrix supporting the growth of the microbial population. In no-till agricultural systems, microbial activities drastically differed with depth, with the greatest microbial activity occurring near the no-till surface; in tilled systems, however, activities were more evenly distributed throughout the plow layer (Doran, 1980). Numbers of all microbial populations usually decrease with increasing depths of soils (Konopka & Turco, 1991), with fungal populations decreasing more rapidly than bacterial (Holm & Jensen, 1980).

The exact role that the biological portion of soil plays in maintaining a high soil quality is unclear. This insufficiency of definition reflects two underlying problems: (i) a lack of a common set of biological indicators and (ii) a recent realization of the magnitude of diversity and interaction within the biotic portion of soil. We argue that the tools to address the structure and function of the soil microbial community are now available. Using these tools, there is now much greater potential to evaluate the role the soil microbial community plays in maintaining a healthy and productive environment. With the advent and use of numerous novel methods for the ecological study of soil, advances in our understanding of the role and function of soil microorganisms and microbial processes in soil quality should be forthcoming.

Our effort is to provide groundwork for selecting proper indicators to estimate the microbial component of soil quality. Consideration is given herein to the tools needed to make assessments of the microbiological portion of soil quality. In particular, we consider methods to address the size and diversity within the soil microbial community, especially soil bacteria. Assessment of microbial diversity may indicate profound differences in soils with respect to microbial populations and functions. Until recently, measures of ecosystem biological diversity have centered on plants, animals, and insect species with little attention given to microorganisms. Recently developed methods will allow in situ estimations of the microbial community structure.

ECOLOGICAL INDICATORS

The *Random House Dictionary of the English Language* (Flexner, 1983) defines *indicator* as a pointing or directing device. An *ecological indicator* is further defined as a plant or animal that indicates by its presence in a given area, the existence of certain environmental conditions. The USEPA's Environmental Monitoring and Assessment Program (EMAP) renames indicators as ecological resource populations (EMAP Monitor, 1992). In most cases, a representative species is selected, and changes in that species are used as a surrogate for the other biological components of the given system. Many individual or groups of organisms have been used as ecological indicators: these include but are not limited to nematodes (Lambshed et al., 1983), moths (Holloway, 1985), birds (Knapp et al., 1991), and earthworms (Lee, 1985). Holloway and Stork (1991) have listed a number of features that should be used in the selection of good ecological indicators. They include the following criteria: the indicator should show a prompt and accurate response to perturbation; the indicator should reflect some aspect of the functioning of the ecosystem; the indicator should be readily and economically accessible; and the indicator must be universal in distribution yet show individual specificity to temporal or spatial patterns in the environment. These criteria also can serve to test the suitability of any organisms or test chosen as an indicator. A better understanding of microbial processes and microbial community structure is needed before the use of bio-indicators will assist in the establishment of long-term strategies for implementing better management practices.

THE BIOTIC COMPONENT

The soil microbial community is a complex mixture that can contain as many as 10 000 different species in a single gram of soil (Torsvik et al., 1990a, b). Well over 50% of the isolated bacterial species reported in *Bergey's Manual* can be found in soil (May, 1988). The earth as a whole is estimated to support between 5 and 10 million species of plants, animals, and microorganisms. The number of different microbial species thought to exist exceeds 1.8 million (Hawksworth & Mound, 1991). Among microorganisms there are 26 900 algae, 30 800 protozoa, 4760 bacteria, 1000 viruses, and 46 983 fungi that are presently described (Wilson, 1988). However, the concept of species as applied to haploid organisms such as bacteria is distinctly different from that applied to organisms capable of sexual reproduction. Among organisms that reproduce in a sexual manner, a *species* often is defined by an inability to interbreed with others not in that species. However, members of a given species are not homogeneous, a reflection of natural genetic variation. It is only when the differences outweigh the similarities that a new species is distinguishable.

A species contains an aggregate of individuals that share a high degree of phenotypic similarity, as well as a number of dissimilarities that can separate them from other groups (Stanier et al., 1976). At this level, the definition of a species is often based on the phenotypic characteristics such as size, shape, etc. Among the microorganisms, it is possible to use phenotypic characteristics to separate species within the algae, protozoa, and fungi; bacteria, because of their common size and structural simplicity, are the exception. Separation among bacteria species is based on testing functional characteristics or difference in their DNA rather than on morphology (Stanier et al., 1976). Present species definition is based on the use of, recovery of, and subsequent purification of individual isolates. In soil, the recent discovery that many viable but not culturable microorganisms exist further confounds the interpretation of species number (Byrd et al., 1991). As a result, the role of these organisms in controlling soil quality is undefined.

MICROBIAL INDICATORS

The most obvious attempt at developing microbial indicators for use in environmental assessment are coliform bacteria. The term *coliform bacteria* describes a broad group of aerobic or facultative, gram-negative, non-spore forming rods that reside in the intestinal tract of all vertebrates and ferment lactose to gas within 48 h when incubated at 35 °C. By definition, coliforms are comprehensive to the genera Escherichia, Klebsiella, and *Enterobacter* (Brock & Madigan, 1991). The presence of coliform bacteria in an environmental sample is used as an indicator of the overall sanitary quality of soil and water environments. Use of an indicator such as coliforms, as opposed to the actual disease-causing organisms, is advantageous, because they generally occur at higher frequencies than the pathogens and are simpler to detect. Although coliforms may serve to indicate an overall environmental status of soil, when the criteria of Holloway and Stork (1991) are considered, coliforms fail as an ecological indicator. Coliforms are not a part of the natural soil ecosystem and indicate the need to carefully choose the correct indicators. Few efforts have been aimed directly at developing indicators of the microbial component of soil quality, but it is possible to use some of the existing tools of soil microbiology to evaluate the microbial components.

When we think about soil as a habitat for microbes, we must consider the variability that exists across as well as within the landscape. With depth, soil has been theorized to be composed of a series of linked ecosystems that can be separated into three parts: surface–root zone, the vadose zone, and the saturated zone. Across the landscape, gross distinctions based on soil color slope, vegetation, etc. are more visible, but from the microbial view point, are not well understood. Within the surface–root zone, both macroorganism and microorganism are common. The total biomass size for the microflora (bacteria/fungi) commonly exceeds 500 mg biomass C kg^{-1} soil (Jenkinson & Ladd, 1981). The vadose zone (subsurface) and aquifer systems are unique in that they are dominated only by microorganisms. Subsurface microbial density, activity, and possibly diversity, appear to be correlated with organic matter content and net water recharge at a particular site.

When developing soil quality indicators to assess the microbial status of a soil, the size of the ecosystem needs to be considered. Three pragmatic questions can be asked: What is being measured and why are we measuring it? Are we trying to measure differences between microbes in a soil at one location, or differences in microbial attributes between locations within a landscape? Are we assessing the microbial community as a native feature of the soil, or are we trying to estimate how the microbial community might respond to perturbation? Consideration of these questions determine the method to estimate microbial parameters. Robertson et al. (1988) have shown that spatial characteristics of N transformations, a microbial mediated process, are complex. They showed that N transformations have a great deal of spatial dependence, generally reflected in the surface landscape. In general, sampling distances of less than 1 m were needed to reduce variability. Rao et al. (1986) showed the degradation rates of the agricultural chemicals, metolachlor [2-chloro-N-(2-ethyl-6-methylphenyl)-N-(2-methoxy-1-methylethyl)acetamide] and aldicarb [2-methyl-2(methylthio)propionaldehyde-O-(methylcarbamoyl)-oxime], were spatially variable. And as pointed out by Rao et al. (1986), measures of microbial degradation can only be properly interpreted as a function of a site's intrinsic and extrinsic features.

The vast microbiological component has been ignored as an important aspect of ecosystem functioning, even though microorganisms are highly sensitive to disturbance and perturbation. Ecosystem functioning before and after disturbance is largely governed by soil microbial dynamics. Soil organisms constitute a large dynamic source and sink of nutrients in ecosystems and play a major role in plant litter decomposition and nutrient cycling (Smith & Paul, 1989). When a disturbed system begins to recover, it is not only due to plant recolonization, but also is dependent on favorable soil conditions and enhanced fertility, which soil organisms promote. In addition, microorganisms promote physiochemical changes in the soil, such as the stabilization of soil organic matter, N_2 fixation, and other alterations in soil properties necessary for plant growth. These differences in microbial properties and activities of soils have been reported but are restricted to general ecological enumeration methods and activity levels, which are limited in their ability to adequately describe a particular ecosystem. Soil organisms are one of the most sensitive biological markers available and the most useful for classifying disturbed or contaminated systems since diversity can be affected by minute changes in the ecosystem. Declining soil microbial diversity as a result of environmental stress needs to be examined (OTA, 1987).

Revegetation of disturbed lands such as mine spoils follows a specific reclamation schedule that depends on the microbial population. Evidence from restoration ecology shows the microbial community is constantly changing throughout reclamation process. Spent oil shale soils are commonly characterized by low microbial numbers, less variety in the types of bacteria, and low rates of cellulose decomposition, compared with a control (Segal & Mancinelli, 1987). Soil reclamation by the addition of topsoil and revegetation was found to increase populations of heterotrophic aerobic bacteria, actinomycetes, ammonium oxidizers, Azotobacter, and fungi to levels found in undisturbed soils. Decline in microbial populations soon after soils are reclaimed can be attributed to utilization of soil organic matter (Fresquez et al., 1986). An additional case study is mycorrhizal changes during restoration. Reclamation of degraded lands often begins with nonmycrorhizal plants with the mycorrhizal inoculum and fungal population increasing with each stage (Cundell, 1977; Waaland and Allen, 1987).

Stabilization of dunes in the Indian desert resulted in a 200-fold increase in microbial populations of fungi, bacteria, and actinomycetes. Azotobacter and nitrifying bacteria in stabilized dunes were higher than in unstabilized dunes. Addition of farmyard manure to desert soils increased the enumerated soil microbial populations, especially fungi (Rao & Venkateswarlu, 1981). Only when diversity studies are conducted will the type and magnitude of the ecological disturbance occurring in a reestablished site be understood.

Any observation of microbial processes should only be interpreted in relation to the interactions with the physical and chemical features of the soil. Our exact understanding of the community structure within soil or the subsurface is fairly limited. Visser and Parkinson (1992) have indicated that studies of soil microorganisms as related to soil quality, can be made at one of three organizational levels: population, community, or ecosystem. They argue that a good indicator of the microbial portion of soil quality should integrate all of the soils featured and should be made at the ecosystem level. To provide a better understanding of the role of soil microbiology in soil quality, it is necessary to discuss the methods that best address and estimate microbial form and function in soil. We feel that the methods to assess microbial form and function in soil may often be divided between those used to estimate diversity between ecosystems and those used to estimate community structure within ecosystems. Both approaches provide needed information about the microbial portion of soil quality.

MICROBIAL FORM AND FUNCTION

Many studies on the in vitro status of soil bacteria have investigated the differences between land uses or treatments. These studies have a comparative nature that discriminates between selected treatment or location effects. For example, Jenkinson and Ladd (1981) reported on the variability in biomass size across a number of soils. Cropping system and tillage effects on the size of the soil microbial biomass have been characterized (Doran, 1980; Dick, 1984; Doran et al., 1987) as have the effects of pesticides (Moorman, 1989). However, assessments of these effects in light of the effects on soil quality are absent. Moreover, assessments that stress an understanding of the metabolic status of the microbial population are rare (Anderson & Domsch, 1990).

Biomass Measurements

Microbial biomass measurements have been used to estimate the biological status of soil. Microbial measurements of this type are critical to assessing the status of a soil, because it represents the fraction of the soil responsible for the energy and nutrient cycling and the regulation of organic matter transformations. A number of different static methodologies are available to estimate biomass size. These methods include microscopic direct counts or observations, chemical (muramic acid) content for bacteria (Miller & Cassica, 1970) and chitin (Frankland et al., 1978), or ergosterol (Newell et al., 1988) for fungi. These methods give a snapshot of the population. Direct observation allows an estimation of the morphological types present in a soil. However, the level of activity of the different portions of the population is not established.

Physiological methods for estimation of total microbial biomass are widely used. These include fumigation–incubation (Jenkinson, 1988) and substrate-induced respiratory response (Anderson & Domsch, 1978). Biochemical analyses have also been used for biomass determinations; these include arginine ammonification (Alef et al., 1988) and the amount of ATP present in the soil (Jenkinson & Ladd, 1981). Each method has its own problems and benefits, but each is able to indicate differences among soils (Table 5-1). No one method has emerged that accurately measures the biomass of a given soil. They all are an index of the activity or mass of the population and each has specific advantages and disadvantages (Smith & Paul, 1989). Although these methods may not be absolute, they are capable of indicating differences or changes in the microbial population and have a major role in microbial analyses.

Since the pioneering work in the late 1920s on bacterial lipids, research has centered on the use of lipids in investigations of microbial form and function (Shaw, 1974; Drucker, 1976; Lechevalier, 1977; Moss, 1981). The phospholipid method as developed by White and Frerman (1967) has been used to study the size of the microbial biomass to depths of 200 cm in the profile (Federle, 1988). The approach uses organic solvents to extract lipids from the environmental sample. This method is advantageous, because it can give estimates of both biomass size and community composition. Phospholipid turnover is rapid because materials in dead cells will be readily consumed

Table 5-1. Estimates of soil microbial biomass by various methods.[†]

Field source	Biomass C		
	Chloroform fumigation	Direct count	ATP method
	——— mg kg^{-1} soil ———		
Continuous wheat plus manure, England	560	500	430
Continuous wheat, no manure, England	220	170	170
Calcareous deciduous wood, England	1230	1400	1040
Acid deciduous wood, England	50	300	470
Old grassland, England	3710	2910	--
Secondary rain forest, Nigeria	540	390	--

[†] From Jenkinson and Ladd, 1981.

by the living biomass. Hence, the total phospholipid content gives a reasonable estimation of the living biomass size (Balkwill et al., 1988).

Enzymes

Numerous studies of enzymes and comparisons of various soils of different climate and management have been conducted (Burns, 1978; Ladd, 1985). Enzymes vary in their location in soil and can be components of the living or dead fractions of soil. For example, dehydrogenase activity has been used since 1956 (Lenhard) to assess microbial activity. This enzyme is intracellular and not extracellular, so that there is a correlation between dehydrogenase and oxygen uptake and therefore the activity of the bacterial population. Sparling (1981), on the other hand, found dehydrogenase not correlated with biomass and other activity measurements. Arginine ammonification (Alef et al., 1988) has been proposed as a general indication of biomass activity. Many other enzymes could be considered as indicators of microbial activity (Chapt. 7, this book).

Activity Measurements

Microbial activity can be assessed in a number of ways that indicate the status of either the total community or in some cases specific members of that community. Many approaches were developed to provide an estimation of the total population response and do not provide estimations of diversity within those populations. The shortcoming of these approaches is that they do not put biomass size in relation to activity or function of the population. In terms of soil quality assessments, the qCO_2 approach may provide a method to relate both the size and activity of the microbial biomass. This may provide an estimation of the effects of environmental change on the population (Anderson & Domsch, 1990).

This approach to describing the biological status of soils is derived from Odum's theory of ecosystem succession (Odum, 1969), which states simply that the ratio of total respiration to total biomass decreases with time or succession in an ecosystem. This can be simplified for soil microbial systems by replacing the variables with basal respiration and the size of the microbial biomass. As seen by Anderson and Domsch (1985, 1990), the respiration/biomass (v/b) ratio becomes a metabolic quotient. This relationship between basal respiration and biomass C is called the metabolic quotient, or qCO_2. It is calculated from basal respiration (CO_2-C h^{-1}) per unit microbial biomass C (C_{micr}). This quotient may be a strong indicator of the effects of environmental influences on the microbial population. Anderson and Domsch (1985, 1990) found qCO_2 to be a sensitive indicator of soil microbial reactions to cropping systems (monoculture vs. crop rotation) and temperature regimes (Fig. 5-1). The soils under crop rotation showed a lower qCO_2, implying a more stable and mature system. Insam and Haselwandter (1989) showed the qCO_2 to decrease with soil age, when comparing glacial moraine soils. With reclamation, the qCO_2 increased, and it was concluded that

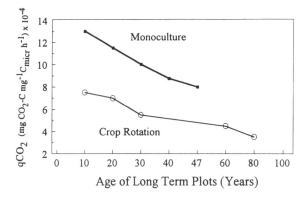

Fig. 5-1. Use of qCO_2 as an indicator of the status of soil microbial biomass (redrawn from Anderson and Domsch, 1990).

qCO_2 values could be related to succession (Insam & Domsch, 1988; Insam & Haselwandter, 1989). This calculation may be a good indicator of the status of a soil and therefore its soil quality.

MICROBIAL DIVERSITY

Efforts to develop methodology for the investigation of microbial diversity have been slow, although some attention has been given to chemically stressed systems such as chemical spills and hazardous waste sites. In these systems, much of the work has focused on specific organisms for potential degradation of toxic substances. Microbial diversity indices have been only minimally used to describe the status of the microbial communities and the effect of natural or human disturbances. Microbial diversity indices can function as bio-indicators by showing the community stability and describing the ecological dynamics of a community and impacts of stress on that community (Mills & Wassell, 1980; Atlas, 1984a, b). An important factor limiting greater use of these indices is the absence of detailed information on the microbial species composition of soil environments.

Culturability—Substrate Utilization

Selection of bacterial isolates from soil is perhaps the most established method of estimating microbial population diversity (Atlas, 1984b). The soil is diluted and plated on specific media and the resulting colonies are identified. In general, the recovered and purified isolate is challenged to use a host of nutrients and biochemical assessments to form a community diversity index (Sorheim et al., 1989; Troussellier & Legendre, 1981; Fredrickson et al., 1991). This method is based on the premise that changes that occur in the microbial community are expressed in the populations that can be cultured. The diversity of the recovered isolates depends, to some degree, on the ex-

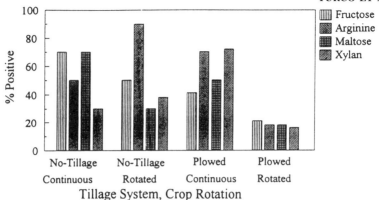

Fig. 5-2. Effect of tillage and crop rotation on substrate utilization by selected bacteria (A.C. Kennedy, 1992, unpublished).

traction method and the type of selection medium utilized (Jensen, 1968; Sorheim et al., 1989). Moreover, it is now clear many soil organisms are viable but not culturable using our present isolation methods (Troussellier & Legendre, 1981; Roszak & Colwell, 1987). Torsvik et al (1990a, b) indicate that 99.5 to 99.9% of the soil bacteria, observed with a flourescence microscope, cannot be isolated or cultured on laboratory media.

Cell maintenance and growth requires energy, C, and numerous inorganic ions; however, species differ in the required compounds needed for metabolic activity. These differences have been used since the beginning of microbiology to distinguish microbes (Hugh & Leifson, 1953; McKinley et al., 1982; Krieg & Holt, 1984; Mergaert et al., 1984; Johnson et al., 1989). Even if only a subportion of the total microbial population grows on the media provided, theoretically it may still provide an indication of how the overall population in that soil is responding to a given use history. For example, the unpublished work of Kennedy (1992) shows that substrate utilization patterns and stress response of microorganisms vary depending on the history of the soil from which they were isolated (Fig. 5-2). Soil microbial populations and activities from soils were evaluated in a 6-yr integrated pest management (IPM) system in the Pacific Northwest. For the groups of microorganisms isolated from the different IPM systems, different patterns of C utilization were evident. Bacterial isolates taken from a plowed, 3-yr rotation treatment were less likely to utilize the C substrates and were less able to withstand the imposed stresses than those isolates from the continuous wheat rotation or wheat in no-till. Isolates from the plow, continuous and no-till, rotation plots were able to use a wider number of the C substrates than from any other crop treatment. Utilization of substrates and characteristics such as response to stress are key to the growth, maintenance, and survival of cells. This information collectively can indicate the makeup of the community. Substrate utilization is key to the survival, growth, and competitiveness of microorganisms in any community (Clark, 1965; Veldkamp et al., 1984). The wide range of compounds available for metabolism and the diverse, yet specific, requirements of specific organisms allows us to use substrate utilization as a discriminating means to characterize and identify bacteria (Fahy & Hayward, 1983; Goor et al., 1984; Mergaert et al., 1984).

Like phospholipid methods, the use of BIOLOG redox technology is an adaptation to soil community evaluation of a technique traditionally used to characterize individual isolates. The BIOLOG system is a method to test the response of microbial isolates to 96 different nutrient sources. Nutrients, as well as an indicator, are contained in the wells of a microtiter plate. A diluted cell suspension is transferred into each well and utilization of the nutrient changes the color of the indicator providing a pattern for a given set of isolates. Garland and Mills (1991), using the BIOLOG plates and diluted soil, reported a community-level assay of microbial community structure without recovery of individual isolates. They found differences in C utilization patterns between soils collected at different locations.

Culturability—Other Classification Methods

Recovered and purified isolates lend themselves to many other classification methods. These methods include protein profiles (Kamicker & Brill, 1986), plasmid profiles (Brockman & Bezdicek, 1989; Lennon & DeCicco, 1991), biochemical and metabolic characterization (Stanier et al., 1976; Martin & Travers, 1989; Kuhn et al., 1991), intrinsic antibiotic resistance (IAR) evaluations (Turco & Bezdicek, 1987; Brockman & Bezdicek, 1989), and many other techniques (Lambert et la., 1990). All of these approaches result in a score for the isolate that can be related back to the population and allow some form of diversity measure to be established (Atlas, 1984a, b). The effects of perturbation can be reflected in changes in the diversity scores. For example, the long-term effects of microclimate on the distribution of microorganisms has been reported (Turco & Bezdicek, 1987; Brockman & Bezdicek, 1989). It is clear from these works that long-term temperature and moisture patterns will influence the hill slope microrelief and result in unique distribution of organisms. These distribution patterns were revealed using IAR and plasmid profiling of the population.

Fatty Acid Methyl Esters (FAME)

Development of phospholipid profiles is an evaluation method that can be used on recovered isolates as well as extracted compounds from soil. When used on isolates recovered from soil, the fatty acids profile are characteristic of specific genera and species; thus, they are reliable for species classification (Moss et al., 1974; Miller, 1982; Miller & Berger, 1985; DeBoer & Sasser, 1986; Zelles et al., 1992). Extraction of phospholipids from soils provides an estimate of community structure from intrinsic features of the population and does not require culturing the organisms. It also is possible using phospholipid extraction to determine the relative types of microbes present in an environment (White et al., 1979; Bobbie & White, 1980; Findlay et al., 1989, 1990; Ringelberg et al., 1989). This community structure assessment is based on an analysis of types of phospholipid fatty acid (PLFA) present. Phospholipid fatty acids are a stable component of cell wall structure of most microorganisms. Following solvent extraction of lipids, the polar lipids are subjected to methanolysis, resulting in FAMEs. The purified FAMEs are subjected to gas chromatography. Zelles et al. (1992) applied this approach to study the community structure in soils maintained under eight different crop management systems. They were able to show differences in the structure of the microbial communities, a direct response to the long-term management effects.

Fatty acid determinations are similar to nucleic acid determinations in taxonomic evaluations (Graham et al., 1990) and in general support the bacterial groupings in *Bergey's Manual* (Lechevalier, 1977; Krieg & Holt, 1984). Other lipids such as phospholipids, glycolipids, hydrocarbons, and isoprenoid quinones (Goldfine, 1972; Lechevalier, 1977; Collins & Jones, 1981; Guckert & White, 1986) also can be used in bacterial identification and diversity.

Nucleic Acid Profiles

Community structure analysis based on direct soil DNA extraction holds great promise in resolving many questions about how perturbations may affect populations. Techniques to recover DNA from soil are now widely accepted (Somerville et al., 1989; Torsvik et al., 1990a, b; Tsai & Olson, 1991). The heterogeneity of the DNA recovered from soil is a reflection of community diversity. Torsvik et al., (1990a, b) have used the kinetics of whole population DNA-reassociation to estimate the number of genomes present in soil. The procedure takes advantage of the finding that the reassociation rates of denatured single-stranded DNA can be used to determine the sequence heterogeneity in the sample (Torsvik et al., 1990a). The denatured DNA is allowed to reassociate and the reassociation rate of the DNA depends on the concentration of homologous DNA strands. The more diverse a population, the longer the reassociation time. The use of this method as a measure of microbial response to environmental stress has not been tested.

Bej et al. (1990, 1991), Josephson et al. (1991), and Pillai et al. (1991) argue that low-level non- or poorly-culturable organisms can be difficult to detect within the mass of extracted soil DNA. They propose that polymerase chain reactions (PCR) offer a way to amplify specific fractions of the total DNA. This PCR approach can take two directions: (i) use of specific primers to amplify exact portions of the populations DNA, as has been done for the detection of *E. coli* in soil and water (Bej et al., 1990, 1991; Josephson et al., 1991), or (ii) use of random primers to randomly amplify portions of the DNA (RAPD). It has been proposed that the RAPD approach will provide a method to generate DNA patterns that may be intrinsically representative of a given soil (Bruce et al., 1992; Picard et al., 1992). The RAPD method has been used to characterize differences among bacterial strains (Williams et al., 1990).

Extracted DNA can be used in concert with DNA–DNA hybridization to detect specific genes in soils. The approach has been used primarily to detect genes that control the catabolic degradation of organic compounds (Fredrickson et al., 1988; Holben et al., 1988), but also has been applied to studies of population dynamics (Calabrese et al., 1991). Again, this approach offers a way to assess the occurrence or loss of abilities within a population.

CONCLUSION

We have a variety of methods to assess the microbial status of a soil. However, few attempts have been made to define what should be included in a data set of microbial indicators of soil quality. As a consequence, we lack a complete data set on the microbial portion of soil quality. The microbial biomass methods are established enough for routine inclusions in a soil quality index. The choice of which biomass method depends on the specific objectives, but in general the physiological methods, particularly the fumigation methods, should be suitable in most instances. Determination of qCO_2 also seems warranted, although more data is necessary before routine recommendations can be made.

In our opinion, the ability to make routine assessments of the status of a soils microbial diversity is critical to our understanding of how perturbations impact the soil system. However, these procedures are still in the experimental stage. The challenge ahead is to define the proper data set so that informed management and regulatory decisions can be made. The indicators we choose need to respond to perturbations, accurately assess the functioning of the system, indicate temporal or spatial differences, and be cost effective and relatively rapid. We cannot expect one method to be adequate for all situations, nor does one method meet all of the criteria set forth by Holloway and Stork (1991). Each assay is an indicator of some portion of soil biology and needs to be selected with consideration of the purpose of the assessment. Our task is to identify the minimal number of biological parameters that consider processes as well as the community diversity for the successful estimation of the role of the biotic component in determining soil quality. These methods also will need to be considered in concert with soil physical and chemical measurements.

REFERENCES

Alef, K., T. Beck, L. Zelles, and D. Kleiner. 1988. A comparison of methods to estimate microbial biomass and N-mineralization in agricultural and grassland soils. Soil Biol. Biochem. 20:561–565.

Anderson, J.P.E., and K.H. Domsch. 1978. A physiological method for the quantitative measurement of microbial biomass in soils. Soil Biol. Biochem. 10:215–221.

Anderson, T.H., and K.H. Domsch. 1985. Maintenance requirements of actively metabolizing microbial populations under in situ conditions. Soil. Biol. Biochem. 17:197–203.

Anderson, T.H., and K.H. Domsch. 1990. Application of eco-physiological quotients (qCO_2 and qD) on microbial biomasses from soils of different cropping histories. Soil Biol. Biochem. 22:251–255.

Atlas, R.M. 1984a. Use of microbial diversity measurements to assess enviornmental stress. p. 540–545. *In* M.J. Klug and C.A. Reddy (ed.) Current perspectives in microbial ecology. Am. Soc. for Microbiol., Washington, DC.

Balkwill, D.L., F.R. Leach, J.T. Wilson, J.F. McNabb, and D.C. White. 1988. Equivalence of microbial biomass measures based on membrane lipid and cell wall components, adenosine, triphosphate and direct counts in subsurface aquifer sediments. Microb. Ecol. 16:73–84.

Bej, A.K., J.L. Dicesare, L. Haff, and R.M. Atlas. 1991. Detection of *Escherichia coli* and *Shigella* spp. in water by using the polymerase chain reaction and gene probes for *uid*. Appl. Environ. Microbiol. 57:1013–1017.

Bej, A.K., S.C. McCarty, and R.M. Atlas. 1990. Detection of coliform bacteria and *Escherichia coli* by multiplex polymerase chain reaction: comparison with defined substrate and plating methods for water quality monitoring. Appl. Environ. Microbiol. 56:2429–2432.

Bobbie, R.J., and D.C. White. 1980. Characterization of benthic microbial community structure by high-resolution gas chromatography of fatty acid methyl esters. Appl. Environ. Microbiol. 39:1212–1222.

Bolton, H., L.F. Elliott, R.I. Papendick, and D.F. Bezdicek. 1985. Soil microbial biomass and selected soil enzyme activities: Effects of fertilization and cropping practices. Soil Biol. Biochem. 17:297–302.

Brock, T.D., and M.T. Madigan. 1991. Major microbial disease. p. 556–557. *In* Biology of microorganisms 6th ed. Prentice Hall, Englewood Cliffs, NJ.

Brockman, F.J., and D.F. Bezdicek. 1989. Diversity of serogroups of *Rhizobium leguminosarum* biovar *viceae* in the Palouse region of Eastern Washington as indicated by plasmid profiles, intrinsic antibiotic resistance and topography. Appl. Environ. Microbiol. 55:109–115.

Bruce, K.D., W.D. Hiorns, J.L. Hobman, A.M. Osborn, P. Strike, and D.A. Ritchie. 1992. Amplification of DNA from native populations of soil bacteria using polymerase chain reaction. Appl. Environ. Microbiol. 58:3413–3416.

Burns, R.G. 1978. Soil enzymes. Academic Press, New York.

Byrd, J.J., H.-S. Xu, and R.R. Colwell. 1991. Viable but nonculturable bacteria in drinking water. Appl. Environ. Microbiol. 57:875–878.

Calabrese, V.G.M., W.E. Holben, and A.J. Sexstone. 1991. Assessment of diversity of 2,4-D-catabolic plasmids in bacteria isolated from different soils. p. 260. *In* Agronomy abstracts. ASA, Madison, WI.

Clark, F.E. 1965. The concept of competition in microbial ecology. p. 339–344. *In* K.F. Baker and W.C. Snyder (ed.) Ecology of soil-borne plant pathogens. Univ. of California Press, Berkeley, CA.

Collins, M.D., and D. Jones. 1981. Distribution of isoprenoid quinone structural types in bacteria and their taxonomic implication. Microbiol. Rev. 45:316–354.

Cundell, A.M. 1977. The role of microorganisms in the revegetation of strip-mined land in the western United States. J. Range Manage. 30:229–305.

DeBoer, S.H., and M. Sasser. 1986. Differentiation of *Erwinia carotovora* spp. *carotovora* and *E. carotovora* spp. *atrosptica* on the basis of fatty acid composition. Can. J. Microbiol. 32:796–800.

Dick, W.A. 1984. Influence of long-term tillage and crop rotation combinations on soil enzyme activities. Soil Sci. Soc. Am. J. 48:569–584.

Doran, J.W. 1980. Soil microbial and biochemical changes associated with reduced tillage. Soil Sci. Soc. Am. J. 44:765–771.

Doran, J.W., D.G. Fraser, M.N. Culik, and W.C. Liebhart. 1987. Influence of alternative and conventional agriculture management on soil microbial processes and nitrogen availability. Am. J. Altern. Agric. 2:99–106.

Drucker, D.B. 1976. Gas-liquid chromatographic chemotaxonomy. Methods Microbiol. 9:51–125.

EMAP Monitor. 1992. Design of comprehensive monitoring program. USEPA Rep. 600/M-91-051. USEPA, Washington, DC.

Fahy, P.C., and A.C. Hayward. 1983. Media and methods for isolation and diagnostic tests. *In* P.G. Fahy and A.C. Hayward (ed.) p. 337–378. Plant and bacterial diseases: A diagnostic guide. Academic Press, New York.

Federle, T.W. 1988. Mineralization of monosubstituted aromatic compounds in unsaturated and saturated subsurface soils. Can. J. Microbiol. 34:1037–1042.

Findlay, R.H., G. King, and L. Watling. 1989. Efficacy of phospholipid analysis in determining microbial biomass in sediments. Appl. Environ. Microbiol. 55:2888–2893.

Findlay, R.H., M.B. Trexler, J.B. Guckert, and D.C. White. 1990. Laboratory study of disturbance in marine sediments: Response of microbial community. Marine Ecol. Prog. Ser. 62:121-133.

Flexner, S.B. (ed.). 1983. The Random House dictionary of English language. 2nd ed. Random House, New York.

Frankland, J.C., D.K. Lindley, and M.J. Swift. 1978. A comparison of two methods for the estimation of mycelial biomass in leaf litter. Soil Biol. Biochem. 10:323-333.

Fredrickson, J.K., D.L. Balkwill, J.M. Zachara, S.M.W. Li, F.J. Brockman, and M.A. Simmons. 1991. Physiological diversity and distributions of heterotrophic bacteria in deep cretaceous sediments of the Atlantic coastal plain. Appl. Environ. Microbiol. 57:402-411.

Frederickson, J.K., D.F. Bezdicek, F.E. Brockman, and S.W. Li. 1988. Enumeration of *Tn5* mutant bacteria in soil by most-probable-number DNA hybridization procedure and antibiotic resistance. Appl. Environ. Microbiol. 54:446-453.

Fresquez, R.R., E.F. Aldon, and W.C. Lindemann. 1986. Microbial diversity of fungal genera in reclaimed coal mine spoils and soils. Reclam. Reveg. Res. 4(3):245-258.

Garland, J.L., and A.L. Mills. 1991. Classification and characterization of heterotrophic microbial communities on the basis of patterns of community-level-sole-carbon-source utilization. Appl. Environ. Microbiol. 57:2351-2359.

Goldfine, H. 1972. Comparative aspects of bacterial lipids. p. 1-51. *In* A.H. Rose and D.W. Tempest (ed.) Advances in microbial physiology. Academic Press, New York.

Goor, M., J. Mergaert, L. Verdonck, C. Ryckaert, R. Vantomme, J. Swings, K. Kersters, and J. De Ley. 1984. The use of API systems in the identification of phytopathogenic bacteria. Med. Fac. Landbouwwetl. Rijksuniv. Gent. 49:499-507.

Graham, J.H., J.S. Hartung, R.E. Stall, and A.R. Chase. 1990. Pathological restriction-fragment length polymorphism, and fatty acid profile relationships between *Xanthomonas campestris* from citrus and noncitrus hosts. Phytopathol. 80:829-836.

Guckert, J.B., and D.C. White. 1986. Phospholipid, ester-linked fatty acid analysis in microbial ecology. p. 455-459. *in* F. Megusar and M. Gantar (ed.) Perspectives in microbial ecology. Proc. of the 4th Int. Symp. of Microbial Ecology, Ljubljana, Yugoslavia. 24-29 Aug. 1986.

Hawksworth, D.L., and L.A. Mound. 1991. Biodiversity database: The crucial significance of collections. p. 17-31. *In* D.L. Hawksworth (ed.) The biodiversity of microorganism and invertebrates: Its role in sustainable agriculture. CAB International, Wallingford Oxon, UK.

Holben, W.E., J.K. Jansson, B.K. Chelm, and J.M. Tiedje. 1988. DNA probe method for the detection of specific microorganisms in the soil bacterial community. Appl. Environ. Microbiol. 54:703-711.

Holloway, J.D. 1985. Moths as indicator organisms for categorizing rain forest and monitoring changes and regeneration processes. p. 235-242. *In* A.C. Chadwick and S.L. Sutton (ed.) Tropical rain-forest. Leeds Philosophical and Literary Society, Leeds.

Holloway, J.D., and N.E. Stork. 1991. The dimension of biodiversity: The use of invertebrates as indicators of human impact. p. 37-63. *In* D.L. Hawksworth (ed.) The biodiversity of microorganism and invertebrates: Its role in sustainable agriculture. CAB International, Wallingford, Oxon, UK.

Holm, E., and V. Jensen. 1980. Microfungi of a Danish beech forest. Holarctic Ecol. 3(1):19-25.

Hugh, R., and E. Leifson. 1953. The taxonomic significance of fermentative versus oxidative metabolism of carbohydrates by various Gram bacteria. J. Bacteriol. 66:24-26.

Insam, H., and K.H. Domsch. 1988. Relationship between soil organic carbon and microbial biomass on chronosequences of reclamation sites. Microbiol. Ecol. 15:177-188.

Insam, H., and K. Hasselwandter. 1989. Metabolic quotient of the soil microfloral in relation to plant succession. Oecologia 79:174-178.

Jenkinson, D.S. 1988. Determination of microbial carbon and nitrogen in soil. p. 368-386. *In* J.B. Wilson (ed.) Advances in nitrogen cycling. CAB International, Wallingford, England.

Jenkinson, D.S., and J.M. Ladd. 1981. Microbial biomass in soil: Movement and turnover. p. 415-471. *In* E.A. Paul and J.M. Ladd (ed.) Soil biochemistry. Marcel Dekker, New York.

Jensen, V. 1968. The plate count technique. p. 158-170. *In* T.R.G. Gray and D. Parkinson (ed.) The ecology of soil bacteria. University Press, Liverpool, England.

Johnson, K.G., M.C. Silva, C.R. MacKenzie, H. Schneider, and J.D. Fontana. 1989. Microbial degradation of hemicellulosic materials. Appl. Biochem. Biotech. 20:245-258.

Josephson, K.L., S.D. Pillai, J. Way, C.P. Gerba, and I.L. Pepper. 1991. Fecal coliforms in soil detected by polymerase chain reaction and DNA-DNA hybridizaiton. Siol Sci. Soc. Am. J. 55:1326-1332.

Kamicker, B.J., and W.J. Brill. 1986. Identification of *Bradyrhizobium japonicum* nodule isolates from Wisconsin soybean farms. Appl. Environ. Microbiol. 51:487–492.

Knapp. C.M., D.R. Marmorek, J.P. Baker, K.W. Thornton, and J.M. Klopatek. 1991. Indicator development strategy for the environmental monitoring and assessment program. USEPA Rep. 600/3-91/023. USEPA, Corvallis, OR.

Konopka, A., and R.F. Turco. 1991. Biodegradation of organic compounds in vadose zone and aquifer sediments. Appl. Environ. Microbiol. 57:2260–2268.

Krieg, N.R., and J.G. Holt. 1984. Bergey's manual of systematic bacteriology. Vol. 1. Williams & Wilkens, Baltimore, MD.

Kuhn, I., G. Allestam, T.A. Stenstrom, and R. Mollby. 1991. Biochemical fingerprinting of water coliform bacteria, a new method for measuring phenotypic diversity and for comparing different bacterial populations. Appl. Environ. Microbiol. 57:3171–3177.

Ladd, J.N. 1985. Soil enzymes. p. 175–221. *In* D. Vaughan and R.E. Malcolm (ed.) Soil organic matter and biological activity. Nijhoff and Junk Publ., The Hague, the Netherlands.

Lambert, B., P. Meire, H. Joos, P. Lens, and J. Swings. 1990. Fast-growing, aerobic, heterotrophic bacteria from the rhizosphere of young sugar beet plants. Appl. Environ. Microbiol. 56:3375–3381.

Lambshed, P.J.D., H.M. Platt, and K.M. Shaw. 1983. The detection of differences among assemblages of marine benthic species based on an assessment of dominance and diversity. J. Nat. Hist. 17:859–874.

Lechevalier, M.P. 1977. Lipids in bacterial taxonomy—a taxonomist's view. Crit. Rev. Microbiol. 5:109–210.

Lee, K.E. 1985. Earthworms: Their ecology and relationships with soils and land use. Academic Press, New York.

Lee, K.E. 1991. The diversity of soil organisms. p. 72–89. *In* D.L. Hawksworth (ed.) The biodiversity of microorganisms and invertebrates; Its role in sustainable agriculture. CASA-FA Rep. Ser. 4. Redwood Press Ltd., England.

Lenhard, G. 1956. Die dehydrogenase—activitat des Bodens als mass für mikroorganismentätigkeit im Boden. Z. Pflanzenernaehr. Bodenkd. 73:1–11.

Lennon, E., and B.T. DeCicco. 1991. Plasmids of *Pseudomonas cepacia* strains of diverse origins. Appl. Environ. Microbiol. 57:2345–2350.

Lipman, J.G., H. McLean, and H.C. Lint. 1916. Sulphur oxidation in soils and its effect on the availability of mineral phosphates. Soil Sci. 1:533–539.

Martin, P.A.W., and R.S. Travers. 1989. Worldwide abundance and distribution of *Bacillus thuringienisis* isolates. Appl. Eviron. Microbiol. 55:2437–2442.

Martynuik, S. and G.H. Wagner. 1978. Quantitative and qualitative examination of soil microflora associated with different management systems. Soil Sci. 125:343–340.

May, R.M. 1988. How many species are there on earth? Sci. 241:1441–1449.

McKinley, V.L., T.W. Federle, and J.R. Vestal. 1982. Effects of hydrocarbons on the plant litter microbiota of an Arctic lake. Appl. Environ. Microbiol. 43:129–135.

Mergaert, J., L. Verdonck, K. Kersters, J. Swings, J.M. Boeufgras, and J. De Lay. 1984. Numerical taxonomy of *Erwinia* species using API systems. J. Gen. Microbiol. 130:1893–1910.

Miller, L.T. 1982. A single derivitization method for bacterial fatty acid methyl esters including hydroxy acids. J. Clin. Microbiol. 16:584–586.

Miller, L.T., adn T. Berger. 1985. Bacteria identification by gas chromatography of whole cell fatty acids. p. 228–241. MIDI Tech. Note 101. MIDI, Newark, DE.

Miller, W.N., and L.E. Casida, Jr. 1970. Evidence for muramic acid in soil. Can. J. Microbiol. 16:299–304.

Moorman, T.B. 1989. A review of pesticides effects on microorganisms and microbial processes related to soil fertility. J. Prod. Agric. 2:14–23.

Moss, C., M.A. Lambert, and W.H. Merwin. 1974. Comparison of rapid methods for analysis of bacterial fatty acids. Appl. Microbiol. 28:80–85.

Moss, C.W. 1981. Gas-liquid chromatography as an analytical tool in microbiology. J. Chrom. 203:337–347.

Mukherjee, D., and A.C. Gaur. 1980. A study of the influence of straw incorporation on soil organic matter maintenance, nutrient release and asymbiotic nitrogen fixation. Zentralbl. Bakteriol. Parasitenkd. Infekttionshr. Hyg. II 135(8):663–668.

Newell, S.Y., T.L. Arsuffi, and R.D. Fallon. 1988. Fundamental procedures for determining ergosterol content of decaying plant material by liquid chromatography. Appl. Environ. Microbiol. 54:1876–1879.

Odum, E.P. 1969. The strategy of ecosystem development. Science (Washington, DC) 164:262-270.

Office of Technology Assessment. 1987. Technologies to maintain biological diversity. OTA, Washington, DC.

Picard, C., C. Ponsonnet, E. Paget, X. Nesme, and P. Simonet. 1992. Detection and enumeration of bacteria in soil by direct DNA extraction and polymerase chain reaction. Appl. Environ. Microbiol. 58:2717-2722.

Pillai, S.D., K.L. Josephson, R.L. bailey, C.P. Gerba, and I.L. Pepper. 1991. Rapid method for processing soil samples for polymerase chain reaction amplification of specific gene sequences. Appl. Environ. Microbiol. 57:2283-2286.

Powlson, D.S., P.C. Brookes, and B.T. Christensen. 1987. Measurement of soil microbial biomass provides an early indication of changes in total soil organic matter due to straw incorporation. Soil Biol. Biochem. 19:159-164.

Ramsay, A.J., R.E. Stannard, and G.J. Churchman. 1986. Effect of conversion from ryegrass pasture to wheat cropping on aggregation and bacterial populations in a silt loam soil in New Zealand. Aust. J. Soil Res. 24:253-264.

Rao, C.R., and B. Venkateswarlu. 1981. Distribution of microorganisms in stabilized and unstabilized sand dunes of the Indian desert. J. Arid Environ. 4:203-207.

Rao, P.S.C., K.S.V. Edvardsson, L.T. Ou, R.E. Jessup, P. Nkedi-Kizza, and A.G. Hornsby. 1986. Spatial variability of pesticide sorption and degradation parameters. p. 100-115. In W.Y. Garner et al. (ed.) Evaluation of pesticide in groundwater. Am. Chem. Soc., Washington, DC.

Ringelberg, D.B., J.D. Davis, G.A. Smith, S.M. Pfiffner, P.D. Nichols, J.S. Nickels, J.M. Hensen, J.T. Wilson, M. Yates, D.H. Kampbell, H.W. Reed, T.T. Stockdale, and D.C. White. 1989. Validation of signature polarlipid fattya cid biomarkers for alkane-utilizing bacteria in soils and subsurface sediments of aquifer materials. FEMS Microb. Eco. 62:39-50.

Robertson, G.P., M.A. Huston, F.C. Evans, and J.M. Tiedje. 1988. Spatial variability in a successional plant community patterns of nitrogen availability. Ecol. 69:1517-1524.

Roszak, D.B., and R.R. Colwell. 1987. Survival strategies of bacteria in the natural environment. Microbiol. Rev. 51:365-379.

Segal, W., and R.L. Mancinelli. 1987. Extent of regeneration of the microbial community in reclaimed spent oil shale land. J. Environ. Qual. 16:44-47.

Shaw, N. 1974. Lipid composition as a guide to the classification of bacteria. Adv. Appl. Microbiol. 17:63.

Smith, J.L., and E.A. Paul. 1989. The significance of soil microbial biomass estimations in soil. Soil biochemistry. Vol. 6. Marcel Dekker, New York.

Sommerville, C.C., I.T. Knight, W.L. Straube, and R.R. Cowell. 1989. Simple, rapid method for direct isolation of nucleic acids form aquitic environments. Appl. Environ. Microbiol. 55:548-554.

Sorheim, R., V.L. Torsvik, and J. Goksoyr. 1989. Phenotypical divergence between populations of soil bacteria isolated on different media. Microb. Ecol. 17:181-192.

Sparling, G.P. 1981. Heat output of the soil biomass. Soil Biol. Biochem. 13:373-376.

Stanier, R.Y., E.A. Adeleberg, and J. Ingraham. 1976. The microbial world. Prentice-Hall, Englewood Cliffs, NJ.

Torsvik, V., J. Goksoy, and F.L. Daae. 1990a. High diversity in DNA of soil bacteria. Appl. Environ. Microbiol. 56:782-787.

Torsvik, V., K. Salte, R. Sorheim, and J. Goksoyr. 1990b. Comparison of phenotypic diversity and DNA heterogeneity in a population of soil bacteria. Appl. Environ. Microbiol. 56:776-781.

Troussellier, M., and P. Legendre. 1981. A functional evenness index for microbial ecology. Microb. Ecol. 7:283-296.

Tsai, Y., and B.H. Olson. Rapid method for direct extraction of DNA from soil and sediments. Appl. Enviorn. Microbiol. 57:1070-1074.

Turco, R.F., and D.F. Bezdicek. 1987. Diversity within two serogroups of Rhizobium leguminosarum native to soils in the Palouse of Eastern Washington. Ann. Appl. Biol. 111:103-114.

Veldkamp, H., H. Van Gemerden, W. Harder, and H.J. Laanbroek. 1984. Competition among bacteria: An overview. p. 279-290. In M.J. Klug and C.A. Reddy (ed.) Current perspectives in microbial ecology. Am. Soc. for Microbiol., Washington, DC.

Visser, S., and D. Parkinson. 1992. Soil biological criteria as indicators of soil quality: Soil microorganisms. Am. J. Altern. Agric. 7:33-37.

Waaland, M.E., and E.B. Allen. 1987. Relationship between VA mycorrhizal fungi and plant cover following surface mining in Wyoming. J. Range. Manage. 40:272-276.

White, D.C., W.M. Davis, J.S. Nickels, J.S. King, and R.J. Bobbie. 1979. Determination of the sedimentary microbial biomass by extractable lipid phosphate. Oecologia 40:51-62.

White, D.C., and F.E. Frerman. 1967. Extraction, characterization and cellular localization of the lipids of *Staphylococcus aureus*. J. Bacteriol. 94:1854-1857.

Williams, J.G.K., A.R. Kubelik, K.J. Livak, J.A. Rafalski, and S.V. Tingey. 1990. DNA polymorphisms amplified by arbitrary primers are useful as genetic markers. Nucl. Acids Res. 18:6531-6535.

Wilson, E.O. 1988. The current state of biological diversity. p. 3-18. *In* E.O. Wilson (ed.) Biodiversity. National Academy Press, Washington, DC.

Zelles, L., Q.Y. Bai, T. Beck, and F. Beese. 1992. Signature fatty acids in phospholipids and lipopolysaccharides as indicators of microbial biomass and community structure in agricultural soils. Soil Biol. Biochem. 24:317-323.

6 Faunal Indicators of Soil Quality[1]

Dennis R. Linden

USDA-ARS
University of Minnesota
St. Paul, Minnesota

Paul F. Hendrix, David C. Coleman, and Petra C.J. van Vliet

Institute of Ecology
University of Georgia
Athens, Georgia

The utility of soil fauna as indicators of soil quality depends on how soil quality is defined. Using criteria from Larson and Pierce (Chapt. 3, this book) and Doran and Parkin (Chapt. 1, this book), soil quality can be considered as the degree to which a soil can (i) promote biological activity (plant, animal, and microbial); (ii) mediate water flow through the environment, and (iii) maintain environmental quality by acting as a buffer that assimilates organic wastes and ameliorates contaminants. Larson and Pierce (Chapt. 3, this book) emphasize that soil quality must be assessed in terms of the specific functions a soil is expected to perform. Given these considerations, in this chapter we review the influences that soil fauna have on soil functional properties that reflect soil quality, and then we explore the utility of soil fauna as indicators of how well these functions are performed and how changes in soil fauna may indicate changes in soil quality. Finally, we consider the specific case of earthworms as soil quality indicators.

SOIL FAUNA AND SOIL PROCESSES

Several recent reviews have considered the influence of soil fauna on soil processes (Swift et al., 1979; Petersen & Luxton, 1982; Anderson, 1988;

[1] Joint contribution of U.S. Department of Agriculture, Agricultural Research Service, St. Paul, MN and Institute of Ecology, University of Georgia, Athens, GA. This work was partially supported by grant BSR-8818302 from the National Science Foundation and a subcontract of grant 58-43YK-0-0045 from the U.S. Department of Agriculture to the University of Georgia Research Foundation.

Coleman & Hendrix, 1988; Crossley et al., 1989; Lavelle et al., 1989; Hendrix et al., 1990; Lee & Foster, 1991; Curry & Good, 1992; Stork & Eggleton, 1992). Processes involved in nutrient cycling and soil structure are of principal interest here, because they form a basis for the three functional properties of soil defined above.

Soil fauna consist of a diverse group of organisms ranging in size from a few micrometers (protozoa) to several centimeters in diameter (land snails) or over a meter in length (large earthworms). Size relationships among major taxonomic groups of soil microflora and soil fauna are illustrated in Fig. 6-1. In terms of biomass and numbers, soil fauna are mostly invertebrates, although certain vertebrates may have significant effects on soil processes in some situations.

The size of soil fauna and their consequent mode of life, especially their mobility and feeding strategies, determine how they influence soil processes. Table 6-1 summarizes the influence of three size categories of soil fauna on soil processes. Microfauna, organisms < 100 μm in width (Protozoa,

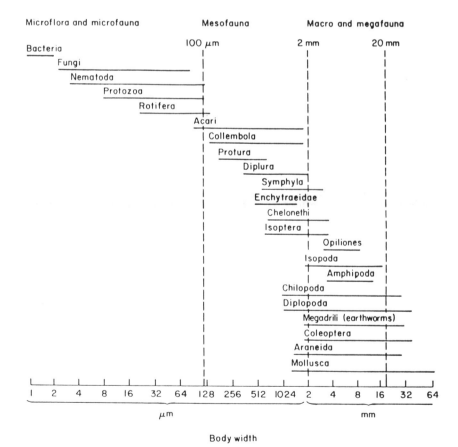

Fig. 6-1. Size classification of soil organisms by body width (from Swift et al., 1979).

Nematoda, and Rotifera), are essentially aquatic organisms that exist in water films on particle surfaces in soil. Because of their small size, they have limited ability to modify soil structure directly. However, they affect soil nutrient availability through their trophic interactions with soil microbes (Elliott et al., 1980; Ingham et al., 1985). Protozoa and certain free-living nematodes consume soil bacteria and fungi. Depending on feeding intensity, these activities may reduce or increase microbial numbers and speed the turnover rate of microbial biomass. The latter effect enhances organic matter mineralization and nutrient availability.

Soil mesofauna, animals 100 to 2000 μm in width, consist of mites (Acari), Collembola, other small insects (Insecta), spiders, (Araneida), and Enchytraeidae, which are in the same class (Oligochaeta) as earthworms. These organisms possess a number of different feeding strategies and functional roles in soil processes. For example, bacterivores and fungivores feed on soil microbes, while omnivores (generalist feeders) and predators feed on other mesofauna. Still others (detritivores) fragment and feed directly on plant residues. Fragmentation increases substrate surface area for further microbial colonization, thereby speeding decomposition and mineralization of organic matter. The larger and more active mesofauna (e.g., enchytraeids) also affect soil porosity through burrowing activities (van Vliet et al., 1993) and aggregation via production of fecal pellets (Rusek, 1985). These processes may increase water infiltration and pore volumes.

Soil macrofauna and megafauna (animals > 2000 μm) are the most conspicuous soil animals and have the greatest potential for direct effects on soil functional properties (Lee & Foster, 1991). These organisms include ants (Formicidae), termites (Isoptera), amphipoda, isopoda, centipedes (Chilopoda), millipedes (Diplododa), adult and larval insects (Insecta), earthworms (Megadrili), snails, and slugs (Mollusca). Some vertebrates (e.g., moles) may

Table 6-1. Influences of soil biota on soil processes in ecosystems.†

	Nutrient cycling	Soil structure
Microflora	Catabolize organic matter Mineralize and immobilize nutrients	Produce organic compounds that bind aggregates Hyphae entangle particles onto aggregates
Microfauna	Regulate bacterial and fungal populations Alter nutrient turnover	May affect aggregate structure through interactions with microflora
Mesofauna	Regulate fungal and microfaunal populations Alter nutrient turnover Fragment plant residues	Produce fecal pellets Create biopores Promote humification
Macrofauna	Fragment plant residues Stimulate microbial activity	Mix organic and mineral particles Redistribute organic matter and microorganisms Create biopores Promote humification Produce fecal pellets

† From Hendrix et al., 1990.

also be important. Soil macrofauna comminute and redistribute organic residues in the soil profile, which increases the surface area and availability of organic substrates for microbial activity. In some soils, these activities may enhance organic matter decomposition and nutrient availability throughout the rooting zone. Certain groups of soil macrofauna, particularly ants, termites, and earthworms, can substantially modify soil structure through formation of macropores and aggregates (Lee & Foster, 1991). These effects may influence infiltration, hydraulic conductivity, and solute leaching through soils (Edwards et al., 1990), and hence a soil's capacity to function as an environmental buffer.

In general, activities of soil animals influence most processes that underlie the functional properties of a soil, i.e., promote plant growth, partition water, and ameliorate environmental contamination. Therefore, some measure of soil faunal abundance, diversity, or activity may provide a useful indicator of soil quality.

SOIL FAUNAL INDICATORS OF SOIL QUALITY

Impacts of human activities on natural assemblages of organisms probably have been recognized since antiquity. However, systematic treatment of altered biotic communities as indicators of environmental change probably began around the turn of this century with the *saprobien system* of Kolkwitz and Marsson (1908, 1909). This was a classification scheme based on their observation that certain organisms are commonly associated with a series of zones of decreasingly severe environmental conditions downstream from organic waste discharges. Lists of species associations were used to define these zones according to the presence or absence of tolerant indicator organisms. Various modifications of this scheme have been used throughout the 1900s to evaluate water quality (e.g., Bick, 1972; Sladecek, 1973).

Until recently, there have been few attempts to use bioindicators to evaluate soil quality (Zlotin, 1985; Foissner, 1987; Paoletti et al., 1991; Stork & Eggleton, 1992). Increasing interest in sustainable agriculture, soil contamination, and soil restoration has resulted in several recent studies of soil fauna as bioindicators. For example, decreasing downwind impacts of point source emissions on soil fauna were shown for industrial chemicals (Ohtonen et al., 1992) and for radioactive fallout from the Chernobyl accident (Krivolutzkii & Pokarzhevskii, 1991). Other studies have examined soil faunal communities in areas contaminated with toxic chemicals compared to noncontaminated areas (Koehler, 1992). Certain species or assemblages of organisms were reported to respond to the disturbance in each case, suggesting their utility as bioindicators of that disturbance. However, the time since disturbance as a factor influencing the response of bioindicators was shown by Beckmann (1988); oribatid mites were *late colonizers* compared with other mesofauna, due to their low dispersal abilities and long developmental time.

Table 6–2 summarizes possible bioindicators of soil quality ranging from single organisms to biological communities and biological processes. At the

Table 6-2. Properties of soil fauna for use as indicators of soil quality.

1. Organisms and populations
   Individuals
      Behavior, morphology, physiology
   Populations
      Numbers and biomass
      Rates of growth, mortality, and reproduction
      Age distribution
2. Communities
   Functional groups
      Guilds (e.g., burrowers vs. nonburrowers, litter vs. soil dwellers, etc.)
   Trophic groups
      Food chains and food webs (microbivores, predators, etc.)
   Biodiversity
      Species richness, dominance, evenness
      Keystone species
3. Biological processes
   Bioaccumulation
      Heavy metals and organic pollutants
   Decomposition
      Fragmentation of organic matter
      Mineralization of C and nutrients
   Soil structure modification
      Burrowing and biopore formation
      Fecal deposition and soil aggregation
      Mixing and redistribution of organic matter

organism level, Curry and Good (1992) suggested that changes in behavior, biochemistry, morphology, physiology, pathology, growth, and reproduction of soil fauna are sensitive measures with *good diagnostic potential*, but the appropriate parameters must be chosen for given site conditions. Changes in relative abundance of species may be good indicators, but it is necessary to show that such changes are due to the disturbance and not to natural temporal fluctuations (Curry & Good, 1992; Koehler, 1992). In general, organism and population level parameters are most sensitive to specific disturbances, perhaps providing *early warnings*, whereas community-level parameters measure disturbance more generally (Mathes & Weidemann, 1990; Curry & Good, 1992).

At the community level, recent studies in applied ecology have addressed the question of soil biological health. Neher et al. (1993) have established sampling protocols to measure nematode populations and community diversity in the Southern Appalachian Piedmont and Coastal Plain regions of North Carolina. This study, part of the USEPA Agroecosystem *E*nvironmental *M*onitoring and *A*ssessment *P*rogram (EMAP) makes use of the fact that nematodes have several attributes that make them useful as ecological indicators (Freckman, 1988; Bongers, 1990):

1. Nematodes are small and have short generation times, so they respond quickly to changes in food supply.
2. Nematodes can survive desiccation and revive with moisture.
3. Populations are relatively stable in soil; thus, any change can be viewed as the result of an environmental perturbation.

4. Perturbations of nematode populations usually reflect a change of trophic structure (Wasilewska, 1991; Coleman et al., 1991).
5. Functional or trophic groups of nematodes can be identified and separated easily (Yeates & Coleman, 1982).

Trophic interactions among various groups of soil fauna result in a cascading or flow of nutrients and energy along pathways that can be recognized as food chains, or, more typically in complex biotic communities, as food webs (Persson et al., 1980; Parker et al., 1984; Ingham et al., 1985). A complex soil food web in a grassland ecosystem is shown in Fig. 6–2. Hunt et al. (1987) recognized both bacterial-based and fungal-based pathways in this model and calculated that the bacterial-based subsystem processed a larger fraction of N flux because of the higher activity of bacteriophagous nematodes and protozoa. The structure and function of soil food webs may serve as useful indicators of management-induced change in soil quality. For example, Hendrix et al. (1986), Holland and Coleman (1987), and Beare et al. (1992) present evidence suggesting greater importance of fungal-based food webs in no-tillage compared with plowed agroecosystems.

Biodiversity indices are often used as measures of change in community structure, but have been used only rarely in studies of soil fauna (Foissner, 1987). Such indices summarize a large amount of information but are costly, time-consuming, and provide little basis for understanding mechanisms underlying changes. Further, diversity indices usually do not provide predictive capabilities or estimates of variability (Curry & Good, 1992).

Graefe (1989, 1990) classified the ecological behavior of earthworms and enchytraeids for soil moisture and soil acidity, similar to the point classification system for vascular plants, developed by Ellenberg (1979). Through this system, the earthworm and enchytraeid communities found at a variety of sites were used as indicators of moisture and acidity conditions at those sites. In different habitats different communities of earthworms and enchytraeids were present. Species composition and abundance within the communities indicated the condition of the sites. Foissner (1987) discovered a similar phenomenon for protozoa. Communities of protozoa only occurred in specific habitats. Foissner (1987) named these groups of protozoa species *eco-orders*.

Soil biological processes may provide useful integrative measures of soil quality. Rates of plant residue decomposition, gaseous evolution from soil (e.g., CO_2, N_2O, CH_4), or soil enzyme activities may be considered indicators of changes in soil fauna, because they are primary microbial reactions that are directly or indirectly influenced by soil fauna. Because soil process rates reflect a complex of activities, observed changes represent a system level response that may affect one or more of the three quality performance functions (promote biological activity, partition water, and ameliorate environmental contamination) of soil. More detailed studies are necessary when cause–effect mechanisms need to be identified.

The bioaccumulation concept as an indicator of soil condition can be illustrated by the example of Thompson and Edwards (1974). Insecticide con-

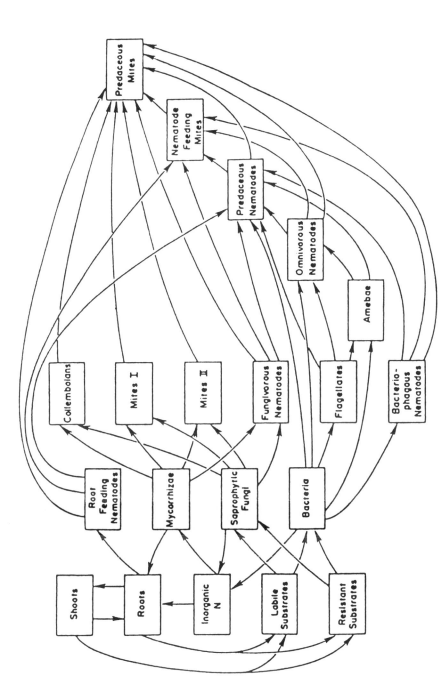

Fig. 6-2. Representation of detrital food web in shortgrass prairie. Fungal-feeding mites are separated into two groups (I and II) to distinguish the slow-growing cryptostimatids from faster-growing taxa. Flows omitted from the figure for the sake of clarity include transfers from every organism to the substrate pools (death) and transfers from every animal to the substrate pools (defecation) and to inorganic N (ammonification) (from Hunt et al., 1987).

centrations in earthworm and slug tissue were shown to be highly correlated with soil insecticide residue concentration. The condition of the soil is thus indicated by the process of accumulation in faunal tissue.

The disappearance, decomposition, and the release of nutrients from crop and animal remains is perhaps one of the most documented processes that is influenced strongly by soil fauna (Wolters, 1991). The organisms, the processes, and the products have all been used to indicate the performance quality of a soil.

Influences of soil fauna on structural properties of soil may be the best long-term indicators of soil quality. For example, the presence of biopores and aggregates formed by macrofauna (earthworms, ants, termites) and the deep mixing of organic and mineral components of a soil (Buntley & Papendick, 1960; Lee & Foster, 1991; West et al., 1991) have a pronounced effect on the functional properties of a soil. Fecal depositions from all levels of soil fauna have been credited with being strong indicators of long-term soil quality (Rusek, 1985; Stork & Eggleton, 1992).

EARTHWORMS AND SOIL PROCESSES

Earthworms are often conspicuous in agriculture and have been credited with indicating high soil quality. Earthworms are perhaps the most studied of the soil fauna, and therefore considerable information is available concerning their potential as bioindicators. Almost every farmer wants to promote earthworm activity because farmers believe earthworms to be beneficial. Literature abounds with earthworm accolades, but is sometimes lacking in definitive comparisons to nonearthworm controls (Darwin, 1881; Hopp & Slater, 1948; Barley, 1961; Lavelle, 1988; Edwards, 1989; Kladivko & Timmenga, 1990; Lee & Foster, 1991; Logsdon & Linden, 1992; Hendrix et al., 1992). As an example organism that has been extensively studied, we will discuss the role of earthworms in modifying the soil environment and the use of earthworms as indicator organisms.

A review of literature on associations between earthworms and properties of soils relevant to the three functions by which soil quality is defined revealed that earthworms contribute to several properties and processes in soils. Earthworms affect the soil environment through their burrowing, fecal excretion (casts), and feeding and digestion. Several excellent review articles have treated this subject (Barley, 1961; Lee, 1983; Daugbjerg et al., 1988; Lavelle, 1988; Edwards, 1989; Kladivko & Timmenga, 1990; Lal, 1991; Lee & Foster, 1991; Logsdon & Linden, 1992).

Earthworm burrows result in increased infiltration capacity and better aeration status of a soil. Macropores created by earthworms are usually many times larger in diameter than the remaining soil pore network and are capable of transporting both air and water at high flow rates. Infiltration rates many times higher than the unburrowed matrix are often reported (Teotia et al., 1950; Bezborodov & Khalbayeva, 1985). Chemical transport during water flow can also be affected by the very fast flux rates associated with these burrows (Edwards et al., 1990; Trojan & Linden, 1992). Surface-applied chemicals can be transported deeper, because they have little opportunity to be adsorbed by soil, whereas chemicals in the soil matrix will move less because flow in macropores bypasses the soil matrix. Large pores drain quickly, and soil aeration after wetting occurs rapidly, which helps keep the soil environment well oxygenated. Rapid air movement in burrows also promotes accelerated soil drying along burrow walls, which helps keep soil well oxygenated.

Earthworm burrows also provide pathways for root exploration of the soil mass (Wang et al., 1986; Jakobsen & Dexter, 1988; Logsdon & Linden, 1992). Pathways created by burrowing are often invaded by roots. In soils with dense or compact horizons, this pathway may be the only route for root penetration into the lower soil horizons (Wang et al., 1986; Logsdon & Allmaras, 1991). Also, burrow walls are nutrient-rich, which may promote root growth and provide a mechanism for enhanced nutrient uptake by crops (Kladivko & Timmenga, 1990). Conversely, earthworm burrowing may, and often does, follow old root pathways. Thus, burrowing and root growth are mutually beneficial activities.

Earthworms ingest, transport, and excrete both organic and mineral matter, thereby affecting soil properties (Lavelle, 1988; Lee & Foster, 1991). The earliest studies of earthworms involved residue (leaf) movement into the soil (Darwin, 1881). Some species of earthworms forage on the soil surface and pull residues into their burrows for later consumption, thereby enhancing rates of residue breakdown and nutrient release. Soil is also mixed by earthworms. High organic matter soil from near the surface often can be seen as excrement in lighter colored subsoil horizons (Nielson & Hole, 1964). Likewise, subsoil can be deposited on, or close to, the surface. This mixing action results in less stratification within the profile (Buntley & Papendick, 1960). Direct and indirect movement of soil minerals by earthworms has been credited with a better mixed chemical environment (Springett, 1983).

Burrow walls and deposited fecal material (casts) are an enriched environment for microbial activity (Daniel & Anderson, 1992). Casts are nutrient-rich and are an intimate mixture of soil, water, and microbial cells. Casts are also usually well-aerated, except when freshly deposited. Reactions such as N_2 fixation, denitrification, and mineralization often proceed at accelerated rates in these locations (Haimi & Huhta, 1990). Accelerated denitrification occurs from freshly deposited casts, and may result in some N loss from agricultural systems (Elliott et al., 1990).

Direct chemical benefits from earthworms have also been reported. Growth-stimulating hormones and acid neutralization are two of these benefits. Acid decomposition products are neutralized during gut passage due to Ca^{2+} release from calcite (Lunt & Jacobson, 1944). Hormones have also been detected in fecal material, but their source is unknown (Tomati et al., 1988).

An important contribution of earthworms to soil environments is the conversion of plant residues into soil organic matter (Lee, 1985; Lavelle, 1988). Some species of earthworms will forage and move residues into soil, which increases residue exposure to microbial activity. Other species feed only within the soil and do not cause large-scale direct displacement of plant residues. Nonetheless, both types of foraging have a major impact on the humification process. Passage of residue material through the gut of earthworms results in physical mixing of soil minerals, water, microbes, and residue matter. Within the gut and after excretion, this material has been shown to be metabolically active (Lunt & Jacobson, 1944; Haimi & Huhta, 1990). Some may argue that earthworms do not directly digest residue material, but rather stimulate the activity of microbes, which may serve as the actual food resource for earthworms. In any case, earthworms ultimately obtain their nutritional requirement from residues and the residue proceeds in the humification process when excreted from the gut. It is this role in the cycling of organic material whereby earthworms may play their most significant role in soils (Barley, 1961; Lavelle, 1988; Lee & Foster, 1991). Nutrient release from organics into the rooting environment means increased opportunity for uptake by crops. While releasing these nutrients, organic material begins to form a structural fabric within the soil (Rusek, 1985; Lee & Foster, 1991).

Earthworms can contribute to erosion and nutrient loss in runoff, because fresh casts are a puddled mass of mineral and organic material. When exposed to rainfall, these casts disperse easily and contribute sediment and nutrients to the runoff stream (Syers et al., 1979). However, as they dry and age, casts generally become more stable than the general mass of soil from which they came, and thereby contribute less to erosion and runoff. Cast stability has been credited to biochemical bonding, humic fabric, and fungal hyphae and mycelium (Shipitalo & Protz, 1989).

Many soil factors, including texture, pH, tillage, residues, and chemical additives, affect the abundance of earthworms (Lee, 1985; Kladivko & Timmenga, 1990). The observation of population change resulting from a practice may lead to the impression that earthworm abundance is an indicator of the practice, but the observations may be site, time, and condition dependent. It has also been suggested that the number of earthworms is not a good indication of their importance (Guild, 1948). Response to a practice may be of only short duration, while long-term effects are ignored. For example, anhydrous ammonia is believed to harm earthworms and nematodes, but it is commonly used to increase crop production. The increased production of a crop is almost always associated with increased above- and below-ground residues that can serve as a food source for soil fauna. Thus, the short-term damage to a population may be outweighed by the long-term abundant food supply.

EARTHWORMS AS INDICATORS OF SOIL QUALITY

A concept wherein earthworms (and perhaps other organisms) can be soil quality indicators is the substrate model analogy in chemistry for the conversion of plant residues into soil organic matter. This model would put earthworm activity as the catalyst facilitating a reaction that produces a product while consuming a substrate. The substrate would be plant residues, which are the basic food source of earthworms, and the product would be soil organic matter. Earthworm activity is limited by the substrate (crop residues) and is thus an indicator of crop residue amounts or the rate of accumulation of soil organic matter. Crop residues have often been reported as the major limitation to a sustained earthworm population (Hopp & Slater, 1948; Teotia et al., 1950; Abbott & Parker, 1981; Jensen, 1985). Crop residues as the precursor for soil organic matter are conceded as one of the most important indicators of soil quality. When earthworms are present and the soil does not have a major limiting chemical or physical condition, then the factor most often limiting to earthworms is food supply. However, there may be upper limits to populations caused by overcrowding, moisture, or other environmental factors that become limiting.

PROBLEMS WITH USING SOIL FAUNA AS INDICATORS

Rigid application of the indicator organism concept has received criticism for several reasons: (i) tolerant organisms can exist in other than disturbed environments; (ii) environmental factors other than disturbance can limit the distribution of species; and (iii) classification of organisms into tolerant, facultative, and intolerant categories is subjective, because tolerances vary under different environmental conditions (Hellawell, 1977).

Furthermore, two critical issues appear often in the literature that bear on the utility of soil fauna as indicators of soil quality. First are the objectives of using bioindicators, which will dictate the appropriate organism, community function, or process to monitor. Although a number of criteria have been suggested for selecting indicators (Day, 1990; Krantzberg, 1990; Curry & Good, 1992), no universally acceptable set of criteria exists. A priori knowledge of site conditions, biological communities, etc. is necessary to make appropriate selections. Therefore, Day (1990) suggested that the first criterion should be *definition of the problem.*

The second issue deals with the soil quality concept. To assess soil quality, some baseline or reference point must be established against which to compare changes in indicator parameters (Mathes & Weidemann, 1990; Curry & Good, 1992). In practice, this baseline will most likely be conditions at specific sites, which will then be monitored through time. Because change in a set of parameters is being measured, Day (1990) argued that "we should be looking at indicators of change, not of 'health' or 'quality,' " which carry unnecessary value judgments. It is clear that soil fauna, like crop responses, provide an integrative measure of soil conditions and of a particular soil's response to disturbance or management.

There are several other difficulties in using earthworms, or perhaps soil fauna generally, as indicators of soil quality. First, their activities in the soil environment are often important but not critical. All of the processes that underlie soil quality (i.e., enhancement of plant growth, partitioning of water, and amelioration of environmental contaminants) can proceed without earthworms. Earthworm activity can in many important instances be considered as a catalyst that enhances the reaction, but not as the only cause of the reaction.

Second, earthworms are often present in highly productive soils but cannot be identified as the cause of the high productivity. Their abundance may be more of a result than a cause, because of large quantities of food resources found in highly productive ecosystems. In other words, there is always a question as to whether earthworm activity is the cause or the effect of high productivity. The answer is, of course, that it is both, because there is a mutually beneficial and synergistic association between earthworm activity and plant production (Lavelle, 1988).

Finally, there are many productive, *high quality* soils that are devoid of earthworm activity. Earthworms may be absent in soils because physical barriers prevent migration or because of unfavorable environmental conditions that humans have, in many cases, overcome through technology (irrigation, drainage, and fertilizer). When introduced, earthworms may thrive in these soils and help maintain soil quality (high productivity and desirable hydrologic properties), but they did not contribute to the high productivity initially.

CONCLUSIONS AND SUMMARY

The large amount of literature reporting measurements of soil properties related to soil quality (e.g., organic matter) and the relatively smaller literature reporting on soil fauna point out a major difficulty in writing this chapter. It is difficult to make broad statements about earthworms, or soil fauna in general, as causes or indicators of high-quality soils. However, stronger associations may be possible between changes in organisms, communities, or processes, and concomitant changes in soil quality functions. Thus, a set of biological, physical, and chemical soil properties should be selected that relate to plant growth, water partitioning, and/or amelioration of environmental contaminants, depending on objectives and local site conditions. Then these properties should be assessed before, during, and after a change in management or imposition of a treatment. Operationally defining the appropriate set of properties for a given situation probably would be more meaningful and useful than attempting to identify an absolute, universal set of indicators.

As a general rule, we suggest that a minimum data set for assessing changes in soil quality should include enumeration and identification of at least the burrowing soil fauna, because of their importance to the structural fabric of a soil, physical and chemical properties of a soil, and their potential contribution to soil quality functions. We would also suggest some measure of faunal fecal depositions (e.g., aggregate-size distribution), which are important to the formation of the soil structural fabric and are thus critical to the long-term quality of a soil. Furthermore, we suggest that the processes of organic matter decomposition and nutrient release, both functions of soil faunal activity, should be assessed because of their influence on nutrient availability and environmental assimilation capacity.

Finally, other potentially useful indicators of change in soil quality under certain conditions include changes in trophic or functional group structure. Nematode and protozoan communities have received most attention, but other assemblages of meso- or macrofauna (e.g., ants) should be considered. Further research in soil biodiversity, both taxonomic and functional, is needed to improve our ability to assess changes in soil quality.

We have reviewed common soil fauna as possible indicators. Soil fauna are conveniently categorized by size into micro-, meso-, and macrofauna. Because of the different functions these groups perform, each size group, or organisms within each group, may serve as indicators of quality depending on the soil functional role of interest. We have reviewed three categories of possible indicators ranging from (i) individual and population levels to (ii) community characteristics including diversity, and trophic associations, to (iii) biological processes including bioaccumulation, and crop residue decomposition. We have reviewed these indicators in terms of three important functions of a soil. A soil's ability to support biological activity, partition water, and ameliorate environmental contaminants are functions upon which the quality factors are assessed. Quality monitoring and maintenance involve one or more of these functions, and soil fauna are strong candidates for indicators of changes in soil quality because of their diverse nature, importance to the soil, and their changes in response to management changes.

REFERENCES

Abbott, I., and C.A. Parker. 1981. Interactions between earthworms and their soil environment. Soil Biol. Biochem. 13:191–198.

Anderson, J.M. 1988. The role of soil fauna in agricultural systems. p. 59–112. In J.R. Wilson (ed.) Advances in nitrogen cycling in agricultural ecosystems. CAB International, Wallingford, England.

Barley, K.P. 1961. The abundance of earthworms in agricultural land and their possible significance in agriculture. Adv. Agron. 13:249–268.

Beare, M.H., R.W. Parmelee, P.F. Hendrix, D.C. Coleman, and D.A. Crossley, Jr. 1992. Microbial and faunal interactions and effects on litter nitrogen and decomposition in agroecosystems. Ecol. Monogr. 62:569–591.

Beckmann, M. 1988. The development of soil mesofauna in a ruderal ecosystem as influenced by reclamation measures: I. Oribatei (Acari). Pedobiologia 31:391–408.

Bezborodov, G.H., and R.H. Khalbayeva. 1985. Effect of the earthworm population on the permeability of sierozems. Sov. Soil Sci. 17:14–17.

Bick, H. 1972. Ciliated protozoa; An illustrated guide to the species used as biological indicators in freshwater biology. World Health Organization, Geneva, Switzerland.

Bongers, T. 1990. The maturity index: An ecological measure of environmental disturbance based on nematode species composition. Oecologia 83:14–19.

Buntley, G.J., and R.I. Papendick. 1960. Worm-worked soils of eastern South Dakota, their morphology and classification. Soil Sci. Soc. Am. Proc. 24:128–132.

Coleman, D.C., A.L. Edwards, A.J. Belsky, and S. Mwonga. 1991. The distribution and abundance of soil nematodes in East African savannas. Biol. Fert. Soils. 12:67–72.

Coleman, D.C., and P.F. Hendrix. 1988. Agroecosystem processes. p. 147–170. In L. Pomeroy and J. Alberts (ed.) Concepts of ecosystem ecology. Springer-Verlag, New York.

Crossley, D.A., Jr., D.C. Coleman, and P.F. Hendrix. 1989. The importance of the fauna in agricultural soils: Research approaches and perspectives. Agric. Ecosyst. Environ. 27:47–55.

Curry, J.P., and J.A. Good. 1992. Soil faunal degradation and restoration. Adv. Soil Sci. 17:171-215.

Darwin, C. 1881. The formation of vegetable mould through the action of worms with observations on their habits. Murray, London.

Daugbjerg, D., J. Hige, J.P. Jenson, and H. Sigurdardottir. 1988. Earthworms as bioindicators of cultivated soils. Ecol. Bull. 39:45-47.

Daniel, O., and J.M. Anderson. 1992. Microbial biomass and activity in contrasting soil materials after passage through the gut of the earthworm *Lumbricus rubellus* Hoffmeister. Soil Biol. Biochem. 24:465-470.

Day, K. 1990. Rapporteur's report of work group: Indicators at the species and biochemical level. Environ. Monit. Assess. 15:277-290.

Edwards, C.A. 1989. Impact of herbicides on soil ecosystems. Crit. Rev. Plant Sci. 8:221-257.

Edwards, W.M., M.J. Shipitalo, L.B. Owens, and L.D. Norton. 1990. Effect of *Lumbricus terrestris L.* burrows on hydrology of continuous no-till corn fields. Geoderma 46:73-84.

Ellenberg, H. 1979. Zeigerwerte der Gefässplanzen Mitteleuropas. 2. Aufl. Scr. Geobot. 9:1-122.

Elliott, E.T., R.V. Anderson, D.C. Coleman, and C.V. Cole. 1980. Habitable pore space and trophic interactions. Oikos 35:327-335.

Elliott, P.W., D. Knight, and J.M. Anderson. 1990. Denitrification in earthworm casts and soil from pastures under different fertilizer and drainage regimes. Soil Biol. Biochem.22:601-605.

Foissner, W. 1987. Soil protozoa: Fundamental problems, ecological significance, adaptations in ciliates and testaceans, bioindicators, and guide to the literature. Prog. Protistology 2:69-212.

Freckman, D.W. 1988. Bacterivorous nematodes and organic matter decomposition. Agric. Ecosyst. Environ. 24:195-217.

Graefe, U. 1989. Zersetzergesellschaften als Standortszeiger-Vorschlag für ein Klassifikationssystem auf der Grundlage von Zootaxozönosen. Verh. Ges. Oekol. 19.1:76.

Graefe, U. 1990. Untersuchungen zum Einfluss von Kompensationskalkung und Bodenbearbeitung auf die Zersetzerfauna in einem boden säuren Buchenwald- und Fichtenforstökosystem. p. 232-241. *In* J. Gehrmann (ed.) Umweltkontrolle am Waldökosystem. Forshung und Beratung, Reihe C, Heft 48, Landwirtschaftsverlag Münster-Hiltrup.

Guild, W.J.M. 1948. Studies on the relationship between earthworms and soil fertility: III. The effect of soil type on the structure of earthworm population. Ann. Appl. Biol. 35:181-192.

Haimi, J., and V. Huhta. 1990. Effects of earthworms on decomposition processes in raw humus forest soil: A microcosm study. Biol. Fert. Soils. 10:178-183.

Hellawell, J. 1977. Biological surveillance and water quality monitoring. p. 69-88. *In* J.S. Alabaster (ed.) Biological monitoring of inland fisheries. Applied Science Publ., London.

Hendrix, P.F., D.A. Crossley, Jr., J.M. Blair, and D.C. Coleman. 1990. Soil biota as components of sustainable agroecosystems. p. 637-654. *In* C.A. Edwards et al. (ed.) Sustainable agricultural systems. Soil and Water Conserv. Soc., Ankeny, IA.

Hendrix, P.F., B.R. Mueller, R.R. Bruce, G.W. Langdale, and R.W. Parmelee. 1992. Abundance and distribution of earthworms in relation to landscape factors on the Georgia Piedmont, USA. Soil Biol. Biochem. 24:1357-1361.

Hendrix, P.F., R.W. Parmelee, D.A. Crossley, Jr., D.C. Coleman, E.P. Odum, and P.M. Groffman. 1986. Detritus food webs in conventional and notillage agroecosystems. BioScience 36:374-380.

Holland, E.A., and D.C. Coleman. 1987. Litter placement effects on microbial communities and organic matter dynamics. Ecology 68:425-433.

Hopp, H., and C.S. Slater. 1948. Influence of earthworms on soil productivity. Soil Sci. 66:421-428.

Hunt, H.W., D.C. Coleman, E.R. Ingham, E.T. Elliott, J.C. Moore, S.L. Rose, C.P.P. Reid, and C.R. Morley. 1987. The detrital food web in a shortgrass prairie. Biol. Fert. Soils. 3:57-68.

Ingham, R.E., J.A. Trofymow, E.R. Ingham, and D.C. Coleman. 1985. Interactions of bacteria, fungi, and their nematode grazers: Effects on nutrient cycling and plant growth. Ecol. Monogr. 55:119-140.

Jakobsen, B.F., and A.R. Dexter. 1988. Influence of biopores on root growth, water uptake and grain yield of wheat (*Triticum aestivum*) based on predictions from a computer model. Biol. Fert. Soils. 6:315-321.

Jensen, M.B. 1985. Interactions between soil invertebrates and straw in arable soil. Pedobiologia 28:59-69.

Kladivko, E.J., and H.J. Timmenga. 1990. Earthworms and agricultural management. p. 192–216. *In* J.E. Box, and L.C. Hammond (ed.) Rhizosphere dynamics. AAAS Selected Symp. 113. AAAS, Washington, DC.

Koehler, H.H. 1992. The use of soil mesofauna for the judgement of chemical impact on ecosystems. Agric. Ecosyst. Environ. 40:193–205.

Kolkwitz, R., and K. Marsson. 1908. Ökologie der pflanzlichen Saprobien. Ber. Dtsch. Bot. Ges. 26A:505–519.

Kolkwitz, R., and K. Marsson. 1909. Ökologie der tierischen Saprobien. Int. Rev. Hydrobiol. Hydrogr. 2:126–152.

Krantzberg, G. 1990. Rapporteur's report of work group: Indicators at community or systems level. Environ. Monit. Assess. 15:283–284.

Krivolutzkii, D.A., and A.,D. Pokarzhevskii. 1991. Soil fauna as bioindicators of biological after-effects of the Chernobyl atomic power station accident. p. 135–141. *In* D.W. Jeffrey and B. Modden (ed.) Bioindicators and environmental management. Academic Press, New York.

Lal, R. 1991. Soil conservation and biodiversity. p. 89–104. *In* D.L. Hawksworth (ed.) The biodiversity of microorganisms and invertebrates: Its role in sustainable agriculture. CAB International, Wallingford, England.

Lavelle, P. 1988. Earthworm activities and the soil system. Biol. Fert. Soils. 6:237–251.

Lavelle, P., I. Barois, A. Martin, Z. Zaidi, and R. Schaeffer. 1989. Management of earthworm populations in agroecosystems: A possible way to maintain soil quality. p. 109–122. *In* M. Clarholm and L. Bergstrom (ed.) Ecology of arable land: Perspectives and challenges. Kluwer Academic Publ., London.

Lee, K.E. 1983. Soil animals and pedological processes. p. 629–644. *In* Soils: An Australian viewpoint. Academic Press, London.

Lee, K.E. 1985. Earthworms: Their ecology and relationships with soils and land use. Academic Press, New York.

Lee, K.E., and R.C. Foster. 1991. Soil fauna and soil structure. Aust. J. Soil Res. 29:745–775.

Logsdon, S.L., and R.R. Allmaras. 1991. Maize and soybean clustering as indicated by root mapping. Plant Soil 131:169–176.

Logsdon, S.L., and D.R. Linden. 1992. Interaction of earthworms with soil physical conditions and plant growth. Soil Sci. 154:330–337.

Lunt, H.A., and G.M. Jacobson. 1944. The chemical composition of earthworm casts. Soil Sci. 58:367–376.

Mathes, K., and G. Weidemann. 1990. A baseline–ecosystem approach to the analysis of ecotoxicological effects. Ecotox. Environ. Saf. 20:197–202.

Neher, D.A., S.L. Peck, J.O. Rawlings, and C.L. Campbell. 1993. Variability of plant parasitic and free-living nematode communities within and between fields and geographic regions of North Carolina. J. Nematol. (in press).

Nielson, G.A., and F.D. Hole. 1964. Earthworms and the development of coprogenous A1-horizons in forest soils of Wisconsin. Soil Sci. Soc. Am. Proc. 28:426–430.

Ohtonen, R., A. Ohtonen, H. Luotonen, and A.M. Markkola. 1992. Enchytraeid and nematode numbers in urban, polluted Scots pine (*Pinus sylvestris*) stands in relation to other soil biological parameters. Biol. Fert. Soils 13:50–54.

Paoletti, M.G., M.R. Favretto, B.R. Stinner, F.F. Purrington, and J.E. Bater. 1991. Invertebrates as bioindicators of soil use. Agric. Ecosyst. Environ. 34:341–362.

Parker, L.W., P.F. Santos, J. Phillips, and W.G. Whitford. 1984. Carbon and nitrogen dynamics during the decomposition of litter and roots of Chihauahuan desert annual *Lepidium lasiocarpum*. Ecol. Monogr. 54:339–360.

Persson, T., E. Baath, M. Clarholm, H. Lundkvist, B.E. Soderstrom, and B. Sohlenius. 1980. Trophic structure, biomass dynamics and carbon metabolism of soil organisms in a scots pine forest. Ecol. Bull. 32:419–459.

Petersen, H., and M. Luxton. 1982. A comparative analysis of soil fauna populations and their role in decomposition processes. Oikos 39:287–388.

Rusek, J. 1985. Soil microstructures—contributions on specific soil organisms. Quaest. Entomol. 21:497–514.

Shipitalo, M.J., and R. Protz. 1989. Chemistry and micromorphology of aggregation in earthworm casts. Geoderma 45:357–374.

Sladecek, V. 1973. System of water quality from the biological point of view. Arch. Hydrobiol., Beih.:Ergeb. Limnol. 7:1–218.

Springett, J.A. 1983. The effect of five species of earthworms on some soil properties. J. Appl. Ecol. 20:865–872.

Stork, N.E., and P. Eggleton. 1992. Invertebrates as determinants and indicators of soil quality. Am. J. Altern. Agric. 7:38–47.

Swift, M.J., O.W. Heal, and J.M. Anderson. 1979. Decomposition in terrestrial ecosystems. Blackwell Scientific Publ., Oxford, England.

Syers, J.K., A.N. Sharpley, and D.R. Keeney. 1979. Cycling of nitrogen by surface-casting earthworms in a pasture ecosystem. Soil Biol. Biochem. 11:181–185.

Teotia, S.P., F.I. Duley, and T.M. McCalla. 1950. Effect of stubble mulching on number and activity of earthworms. Univ. of Nebraska Agric. Exp. Stn. Res. Bull. 165:1–20.

Thompson, A.R., and C.A. Edwards. 1974. Effects of pesticides on non-target invertebrates in freshwater and soil. p. 341–386. *In* W.D. Guenzi (ed.) Pesticides in soil and water. SSSA, Madison, WI.

Tomati, U., A. Grappelli, and E. Galli. 1988. The hormone-like effect of earthworm casts on plant growth. Biol. Fertil. Soils 5:288–294.

Trojan, M.D., and D.R. Linden. 1992. Microrelief and rainfall effects on water and solute movements in earthworm burrows. Soil Sci. Soc. Am. J. 56:727–733.

Van Vliet, P.C.J., L.T. West, P.F. Hendrix, and D.C. Coleman. 1993. The influence of Enchytraeidae (Oligochaeta) on the soil porosity of small microcosms. Geoderma 56:287–299.

Wang, J., J.S. Hesketh, and J.T. Wooley. 1986. Preexisting channels and soybean rooting patterns. Soil Sci. 141:432–437.

Wasilewska, L. 1991. Long-term changes in communities of soil nematodes on fen peat meadows due to the time since their drainage. Ekol. Pol. 39:59–104.

West, L.T., P.F. Hendrix, and R.R. Bruce. 1991. Micromorphic observation of soil alteration by earthworms. Agric. Ecosyst. Environ. 34:363–370.

Wolters, V. 1991. Soil invertebrates—effects on nutrient turnover and soil structure—a review. Z. Pflanzenernaehr. Bodenkd. 154:389–402.

Yeates, G.W., and D.C. Coleman. 1982. Nematodes in decomposition. p. 55–80. *In* D.W. Freckman (ed.) Nematodes in soil ecosystems. Univ. of Texas Press, Austin, TX.

Zlotin, R.I. 1985. Role of soil fauna in the formation of soil properties. Soil Ecol. Manage. 12:39–47.

7 Soil Enzyme Activities as Indicators of Soil Quality[1]

Richard P. Dick

Oregon State University
Corvallis, Oregon

With the increasing pressure to produce more food, fiber, and fuel to meet world demands on a limited land area, there is an unprecedented need to address global concerns about soil degradation. Understanding the underlying biological processes in tandem with identification of early warning indicators of ecosystem stress is needed to provide strategies and approaches for land resource managers and policymakers to promote long-term ecosystem sustainability.

Biological and biochemically mediated processes in soils are fundamental to terrestrial ecosystem function. Ultimately, all members of the food web are dependent on the soil as a source of nutrients, and for degradation and cycling of complex organic compounds. Primary decomposers of organic matter provide energy that supports the activities of organisms from a number of trophic levels in soils.

Historically, chemical and physical properties have been used as crude measures of soil productivity. Most notably, determination of soil organic matter has been related to general soil tilth. Soil organic matter changes very slowly, and therefore, many years may be required to measure changes resulting from perturbations. However, there is growing evidence that soil biological parameters may hold potential as early and sensitive indicators of soil ecological stress or restoration (Dick, 1992; Dick & Tabatabai, 1992). In Chapter 5 (this book) information is provided on soil quality in relation to soil microorganisms. In this chapter, soil enzyme activities will be discussed as a potential biochemical/biological indicator of soil quality.

ORIGINS AND LOCATIONS OF SOIL ENZYMES

It is generally assumed that soil enzymes are largely of microbial origin (Ladd, 1978), but it is also possible that animals and plants may contribute enzymes to soils. It is difficult to conclusively discriminate between sources

[1]Oregon Agric. Exp. Stn. Journal no. 10,195.

of enzymes in soils, and thus, evidence for the primary role of microbes as a source of soil enzymes results from indirect evidence (for a more detailed discussion, see the reviews of Ladd, 1978, and Skujinš, 1978).

Relatively little information is available on the contribution of enzymes to soil from soil fauna. Urease (EC 3.5.1.5) (Syers et al., 1979), phosphatases (EC 3.1.3.-) (Sharpley & Syers, 1977; Park et al., 1990, 1992), and invertase (EC 3.2.1.26) activities (Kiss, 1957) have been reported in earthworm (*Lumbricus terrestris*) casts whereas ants (*Formica*) made little contribution to soil invertase activity (Kiss, 1957). Kozlov (1965) concluded that soil fauna provide only a limited contribution of enzymes to soil.

Vegetation can potentially affect soil enzyme content directly or indirectly. Plant roots can excrete extracellular enzymes (Rogers, 1942; Rogers et al., 1942; Estermann & McLaren, 1961; Dick & Tabatabai, 1986). Skujinš (1978), in his review, and later reports by Speir et al. (1980), and Castellano and Dick (1991) showed that, in general, rhizosphere soil increases enzymatic activity over nonrhizosphere soil for such enzymes as phosphatases, nucleases, invertase, urease, catalase (EC 1.11.1.6), arylsulfatase (EC 3.1.6.1), and protease. These studies, however, could not distinguish the enzymes released by roots from enzymes of microbial origin. It seems likely that roots stimulating microbial activity (Alexander, 1977) would be a major factor for explaining the rhizosphere effect on soil enzyme activity.

Plant residue contains enzymes that may be released into the soil upon residue decomposition or remain active in partially decomposed plant tissue. It is difficult to conclusively separate plant tissue–derived enzymes from that of microbial origin. Indirect evidence presented by Ladd (1978) showed that extracted enzymes, possibly originating from plant debris, were rapidly decomposed and favored microbial growth in incubated soils. Furthermore, Dick et al. (1983) showed that fresh corn (*Zea mays* L.) root tissue exhibiting acid phosphatase activity, when added to soils, was rapidly degraded by soil proteases and/or inhibited by soil constituents. Additions of steam sterilized root homogenate (i.e., inactivated acid phosphatase activity) increased acid phosphatase activity, suggesting that plant residue stimulated microbial synthesis of acid phosphatase rather than providing direct additions of corn root phosphatase to the soil. Nannipieri et al. (1983) found similar results when ryegrass (*Lolium* sp.) tissue at time zero increased phosphatase, urease, and protease, but after 30 to 60 d incubation the activities of these enzymes returned to control levels.

Burns (1982) classified enzymes according to their location in the soil. Three enzyme categories (termed *biotic enzymes*) were associated with viable proliferating cells located: (i) intracellularly in cell cytoplasm, (ii) in the periplasmic space; and (iii) at the outer cell surfaces. The remaining categories are broadly characterized as *abionitic*, a term first used by Skujinš (1976) that is derived from the Greek a-, the alpha privatitive meaning "removal or absence of a quality," and the Greek suffix -biontic, meaning "having a form of life." Abiontic enzymes, then, are those exclusive of live cells that include enzymes: (i) excreted by living cells during cell growth and division, (ii) attached to cell debris and dead cells, and (iii) leaked into soil solution from extant or lysed cells but whose original functional location was on or within the cell. Additionally, abiotic enzymes can exist as stabilized enzymes in two locations: adsorbed to internal or external clay surfaces, and complexed with humic colloids through adsorption, entrapment, or copolymerization during humic matter genesis (Boyd & Mortland, 1990).

The activity of approximately 50 to 60 enzymes have been identified in soils (see review by Ladd [1978] and Skujinš [1967] for a detailed listing of soil enzymes that have been studied). Undoubtedly, the number in soils is far greater, but techniques for determining the presence or activity of other enzymes have not yet been developed for soils. The soil enzymes most often studied are oxidoreductases, transferases, and hydrolases. The oxidoreductase, dehydrogenase, has been widely studied in soils partly because of its apparent role in the oxidation of organic matter where it transfers hydrogen from substrates to acceptors. Although dehydrogenase activity depends on the total metabolic activity of the viable microbial populations, and should exist only in integral parts of intact cells, it has not always reflected total numbers of viable microorganisms isolated on a particular medium or with oxygen consumption or CO_2 evolution (Skujinš, 1976). Catalase activity is based on the rate of release of oxygen from added hydrogen peroxide or the recoveries of hydrogen peroxide.

Some hydrolases and transferases have been extensively studied because of their role in decomposition of various organic compounds, and thus are important in nutrient cycling and formation of soil organic matter. These would include enzymes involved in:

1. The C cycle—amylase (EC 3.2.1.-), cellulase (EC 3.2.14), lipase (EC 3.1.1.3), glucosidases, and invertase
2. The N cycle—proteases, amidases, urease, and deaminases
3. The P cycle—phosphatases
4. The S cycle—arylsulfatase

Lyases have shown activity in soils but relatively few studies have been conducted on this group of enzymes.

ECOLOGY AND FUNCTION OF SOIL ENZYMES

Soils may be considered a biological entity (Quastel, 1946) with complex biochemical reactions. Soil enzymes play an important role in soil microbial ecology by catalyzing innumerable reactions in soils. Enzymes are proteins that act as catalysts without undergoing permanent alteration and cause chemical reactions to proceed at faster rates. Enzymes are specific activators because they combine with their substrates in stereospecific fashion that decreases the stability of certain susceptible bonds (i.e., changes in electronic configuration).

Clearly, intracellular enzymes play an important role in the life processes of microorganisms in the soil. Although a few enzymes can only function within a viable cell (e.g., dehydrogenase), the vast majority of enzymes identified in soil appear to remain functional (if only for short periods) in soil solution, in dead cells, on cell debris, or complexed in the soil matrix.

The functional role of abiontic soil enzymes has yet to be conclusively shown experimentally. First, there has been limited success in extracting enzymes from soils (Tabatabai, 1982). This is because the enzymes complexed in the soil matrix often lose their integrity during the extraction process. As a result, soil enzymes have largely been studied by measuring activities that makes separation of abiontic enzymatic activities indistinguishable from activities associated with the living organisms. Secondly, the degree to which abiontic enzymes catalyze naturally occurring substrates at in situ concentration is not known.

Nonetheless, there is an emerging conceptual model, based on experiments of model humic/clay–enzyme complexes (Paul & McLaren, 1975; Boyd & Mortland, 1990), which suggests abiontic enzymes may have important relationships with soil organisms (Burns, 1982). Burns (1982) hypothesized that humic–enzyme complexes are important for substrate catalysis. For some organisms, substrates or their breakdown products may be useful for the organism, but due to substrate size or insolubility are unavailable for microbial uptake. Therefore, a microorganism may have no means to detect the presence of the substrate, and thus will not be induced to synthesize and secrete the extracellular enzymes to hydrolyze the compound. However, the enzyme associated with a complex would not be under the same regulatory control as it would if it were intercellular and could therefore respond rapidly. In turn, the product released by the complexed soil enzyme could be further broken down by other soil enzymes or be taken up by microorganisms. It may be an advantageous situation for a microbial cell to be located on the surface of a humic colloid containing a number of enzyme molecules. Indeed, for some species, their successful survival in a hostile soil environment may depend on an association with humus–enzyme complexes.

The above discussion suggests that, in terms of soil ecosystem functioning, optimal levels of soil enzymes may be important in questions of soil quality.

ENZYME ACTIVITY IN RELATION TO SOIL BIOLOGY

Since about the 1960s, there have been various attempts to correlate enzyme activity with soil microbial activities. Early work gave mixed results, and in his review, Skujiņš (1967) reports very few direct correlations of activities of various soil enzymes with CO_2 evolution and direct counts of microbes. This would not be expected for several reasons. First there may be inadequacies of the measurement method of microbial activity at that time. In the case of respiration, microorganisms may not be the only source of CO_2, and total microbial counts are not accurate measures of microbial activity, because they are biased by the medium selected for cultivation of microorganisms. Another factor is soil enzyme activity measurement. A given enzyme performs a very specific reaction; therefore, only a small fraction of the population may possess that enzyme at any given time. In addition, the activity of many enzymes in soil is a composite of activities from biotic and abiontic components. Therefore, it might be possible to have a low viable microbial population in a soil that has a high abiontic enzyme content, which would result in a poor correlation between enzyme activity and microbial activity or biomass.

These early studies showed the need to relate enzyme activity to soil microbiology for a specific purpose. As new microbial and enzyme assays became available, stronger relationships between enzyme activities and soil biological parameters have been identified. Frankenberger and Dick (1983) evaluated 11 enzymes in 10 diverse soils for their potential relationships with microbial respiration, biomass, viable plate counts, and other soil properties. They found that alkaline phosphatase (EC 3.1.3.1), amidase, and catalase were highly correlated with both microbial respiration and total biomass in soils but not with microbial plate counts. This indicates that enzyme activities were associated with microorganisms that were active in soil but whose numbers were not accurately reflected by the viable plate counts. Other studies have shown microbial activity to be correlated with the activity of such enzymes as dehydrogenase, protease, cellulase, phosphatase, and urease (Laugesen, 1972; Laugesen & Mikkelsen, 1973; Nannipieri et al., 1978; Ross & Cairns, 1982; Tiwari et al., 1989). Alef et al. (1988) have proposed arginine ammonification as a biological index. They reported this to be a relatively simple assay that showed high correlation with biomass C, heat output, soil ATP content, and soil protease activity ($r > 0.90$), but not with viable microbial counts, in 22 soils from Germany and Austria.

Asmer et al. (1992) creatively showed a close relationship between extracellular protease activity and several biological parameters. In an incubation study, soil amended with Kings agar to induce microbial growth was sampled over a 12-d period and excreted protease was separated from the soil matrix by membrane filtration after extracting with Modified Universal Buffer. The advantage of this approach was that it nondestructively separated protease excreted into soil solution from protease of the soil matrix (previous attempts have used steam sterilization, radiation, or antiseptics; see review of Skujinš [1976] for detailed discussion of these studies). Soluble protease activity was highly correlated with ATP content ($r = 0.74$) and total bacteria counts ($r = 0.94$). Besides showing a relationship between enzyme activity and soil biology, it also provided evidence for a functional role of extracellular enzymes as hypothesized by Burns (1982), i.e., extracellular enzymes begin hydrolyzing compounds outside the cell that may be too large or insoluble for microorganisms to use directly.

To circumvent the problems of enzymes that can exist both extracellularly and intracellularly, attempts have been made to utilize enzymes that could only exist in viable cells. The most widely studied enzyme has been dehydrogenase. However, it has not consistently correlated with microbial activity (Skujinš, 1978; Frankenberger & Dick, 1983). Howard (1972) reviewed the literature and presented data that provided a potential explanation of these results. Using oxygen uptake to calculate theoretical dehydrogenase activity, he showed that the observed dehydrogenase activity was substantially less than theoretical dehydrogenase activity. He hypothesized that extracellular phenol oxidases, which are known to exist in soil, may also carry out the dehydrogenase reaction, thus reducing the potential for dehydrogenase activity to serve as an index of viable microbial activity.

Another relatively untested approach that assays indiscriminate hydrolytic activity is the measurement of the hydrolysis of fluorescein diacetate (3', 6'-diacetylfluorescein) (FDA) (Schnürer & Rosswall, 1982). Fluorescein diacetate is hydrolyzed by a number of different enzymes, such as proteases, lipases, and esterases (Guilbaut & Kramer, 1964; Rotman & Papermaster, 1966). The product of this reaction is fluorescein, which can be quantified by fluorometry or spectrophotometry. Fluorescein diacetate hydrolysis appears to be widespread among the primary decomposers, bacteria, and fungi (Lundgren, 1981; Medzon & Brady, 1969; Söderström, 1977). Perucci (1992) found rates of FDA hydrolysis in soils amended with municipal compost measured over a 3-yr period to be highly correlated with activities of amylase, arylsulfatase, deaminase, and protease, and with microbial biomass C. This assay is not specific to a particular enzyme, which in terms of soil quality measurements, could be an advantage over enzyme assays that measure rates of specific reactions because organic residue–decomposing organisms appear to be the major contributors to soil enzyme activity (Speir, 1977; Speir & Ross, 1976). Thus, FDA hydrolysis should broadly represent an important microbial component that plays a central role in the ecology of soils.

Although it seems unlikely that soil enzymes directly participate in soil structure development, there is evidence that soil biology has an important role. For example, Chesters et al. (1957) showed that additions of organic residues to sterile soils resulted in little improvement in soil structure. Capriel et al. (1990) reported significant correlations between soil aggregate stability and soil microbial biomass. Therefore, if it is true that decomposers are the primary source of soil enzymes, then it is possible that a correlative relationship exists between soil enzyme activities and soil structural parameters. Indeed, the few studies that have examined this relationship have shown correlations between soil enzyme activities and soil structural factors. Dick et al. (1988a), in compacted and uncompacted forest soils, found highly significant negative correlations between bulk density and activities of dehydrogenase, phosphatase, and arylsulfatase. In 7 of 10 enzymes tested, Martens et al. (1992) found significant negative correlations with soil bulk density, and in five enzymes significant positive correlations with cumulative water infiltration rates. A comparison of on-farm cropping systems showed that an alternative system that included a legume green manure had improved soil structure (Reganold, 1988) and increased soil enzyme activities (Bolton et al., 1985) over a conventional system. If this is a consistent relationship, measurement of soil enzyme activities would be a more practical means for indicating changes in soil structure because the procedure is considerably simpler than the labor-intensive procedures of standard soil structural methods.

SOIL ENZYME ACTIVITIES AS INDICATORS
OF PERTURBATION

Soil Fertility Indicator

With the advent of increased numbers of soil enzyme assays in the 1950s, there was considerable interest and effort in Europe and Russia to develop soil enzyme activities to be used as a *soil fertility* index. The goal was to provide a practical tool for agriculture that would complement chemical soil tests and aid in assessing the nutrient status of soils and be correlated with crop yield. The results of these studies gave conflicting results, which was partially due to flawed soil enzyme methodologies (Skujinš, 1978). But in his review, Skujinš (1978) cites work as early as 1954 (Koepf, 1954a, b) that showed no close relationship between soil enzyme activities and nutrient status. Subsequent researchers came to similar conclusions with regard to relating enzyme activities or other biological parameters to crop yields or soil nutrient status (Drobník, 1957; Galstyan, 1960; Hábán, 1967). Conversely Verstraete and Voets (1977) showed that selected soil enzymes (phosphatase, invertase, β-glucosidase [EC 3.2.1.21], and urease) were correlated with crop yields and that these were superior to microbial enumeration in correlating with soil fertility or crop yield.

In unmanaged ecosystems (Skujinš, 1978) or low-input agricultural systems, a stronger relationship between soil enzyme activity or other biological parameters and plant biomass production might be expected. However, in managed systems, other factors may confound or override the relationship between soil biological activity and plant productivity. This would likely be true for agroecosystems where high amounts of external inputs of nutrients and water can greatly stimulate plant growth without a corresponding response by soil microorganisms. Yaroshevich (1966) showed that manure-amended soil increased soil respiration and enzyme activity, but inorganic fertilizer–amended soils showed depressed biological activity. Crop yields, however, were the same when adequate nutrients were supplied from either inorganic or organic sources.

Evidence for the confounding effect of fertilizers is shown in Fig. 7–1. Here, Chunderova and Zubets (1969) showed that after 4-yr of cropping, soils with high solution P concentrations due to increasing rates of P fertilization depressed phosphatase activity. Dick et al. (1988b), on soil samples from plots with treatments that had been in place since 1931, showed a similar effect where increasing rates of ammonia based N fertilizer decreased the activity of amidase and urease (both enzymes that are involved in the N cycle). Conversely, activities of other enzymes not directly involved in the N cycle (arylsulfatase and β-glucosidase) did not correlate with N fertilizer input levels. A feedback mechanism was hypothesized as suppressing production of enzymes whose reaction product (NH_4) was continually added by inorganic fertilizers. Thus, it is important to carefully chose enzymes that might be affected by fertilization practices when considering their use in measuring soil quality.

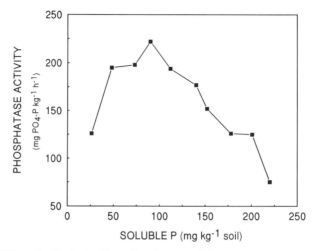

Fig. 7-1. Effect of soil solution P on phosphatase activity (as measured by release of phosphate from glycerol phosphate) in soils from field plots receiving varying rates of inorganic P fertilizer (adapted from Chunderova & Zuberts, 1969).

Long-Term Effects of Soil Management

Studies from long-term sites have shown that soil enzyme activities are sensitive in discriminating among soil management treatment effects. Organic amendments such as animal manures (Yaroshevich, 1966; Khan, 1970; Verstraete & Voets, 1977; Dick et al., 1988b; Martens et al., 1992), green manures/crop residues (Verstraete & Voets, 1977; Dick et al., 1988b; Martens et al., 1992), and municipal refuse (Werner et al., 1988; Perucci, 1992) significantly increased the activity of a wide range of soil enzymes over unamended soil. Although these organic amendments can often contain enzymes, the increase in activity in soils amended with organic residues is likely due to stimulation of microbial activity rather than direct addition of enzymes from the organic sources (Martens et al., 1992).

Comparing crop rotations with monoculture systems has shown that soil enzyme activities are sensitive to the positive effects associated with multicrop systems (Khan, 1970; Blagoveshchenskaya & Danchenko, 1974; Dick, 1984; Bolton et al., 1985).

Soil enzyme activities have been responsive to tillage treatments as well. Gupta and Germida (1988) compared cultivated soils (69 yr) with adjacent grassland and found that cultivation depressed phosphatase (49%) and arylsulfatase (65%) activity. Effects of cultivation on soil biology can be modified by the type of tillage. Conservation tillage practices produce less soil disturbance than conventional tillage systems and have higher levels of enzyme activities in the surface soil (< 10-cm depth) (Klein & Koths, 1980; Doran, 1980; Dick, 1984). As an example, Fig. 7-2 shows a multisite study where dehydrogenase activities increased from 10 to 190% in the 0- to 7.5-cm depth ($P < 0.05$, except at the West Virginia site) in no-till systems over con-

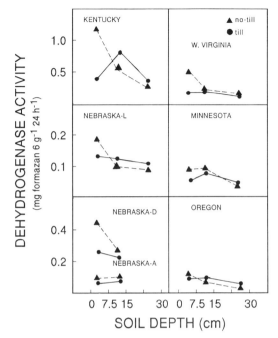

Fig. 7-2. Soil dehydrogenase activity for no-till and conventionally tilled soils from three soil depths (adapted from Doran, 1980).

ventional systems. A similar response was found for phosphatase activity, which was 10 to 150% higher in no-till than conventional tillage.

There is very little information on the relationship between soil perturbations and soil enzyme activity in forest ecosystems. Dick et al. (1988a) compared the effect of various soil tillage treatments applied to compacted soils in a forest clear-cut in western Oregon. Four years after the treatments were initiated, phosphatase activity was significantly decreased in compacted soil from 10 to 60 cm deep compared with rehabilitated soil. All enzymes tested (phosphatase, amidase, dehydrogenase, and arylsulfatase) were depressed from 41 to 75% in the 10- to 20-cm depth. Enzyme activities were negatively correlated with soil bulk density ($P < 0.05$). Biological parameters such as soil enzyme activities could provide forest managers a way to identify management practices that are improving soil properties, long before such benefits could be measured by tree growth.

In some cases, enzyme activity can detect the effects of soil pollutants (see review by Kiss et al., 1975). For example, several studies have shown that heavy metals associated with sewage sludge can inhibit soil enzyme activity (Frankenberger et al., 1983; Bonmati et al., 1985).

Temporal Responsiveness

To be useful, a soil quality index needs to be responsive to soil perturbations or restoration efforts in a relatively short time. Figure 7-3 presents

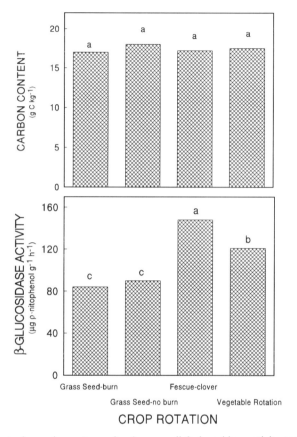

Fig. 7-3. Effect of cropping systems after 2 yr on soil β-glucosidase activity and total C contents in western Oregon. Bars with the same letter are not significantly different ($P < 0.05$) according to Duncan's multiple range test (unpublished, Dick & Miller, 1992).

unpublished data by Dick and Miller (1992) at a replicated site in western Oregon. Two years after imposing various cropping systems, a significant effect on β-glucosidase activity was observed even though there were no measurable differences in total C content. Perucci (1992) found that the activities of eight enzymes were significantly greater within 30 d after municipal treatments were incorporated and higher activities continued to be observed over the following 3 yr (except for dehydrogenase activity). Martens et al. (1992) showed that a range of organic residues (poultry manure, sewage sludge, and plant residues) generally caused a significant increase in activities of key enzymes involved in the C, N, P, and S cycles 30 d after the first application. They hypothesized that a trigger molecule or promoter is released by decomposing organic compounds that stimulates the production of hydrolytic enzymes.

Another important temporal factor for a successful soil index is variability, both within season and season to season. Wide variability would make

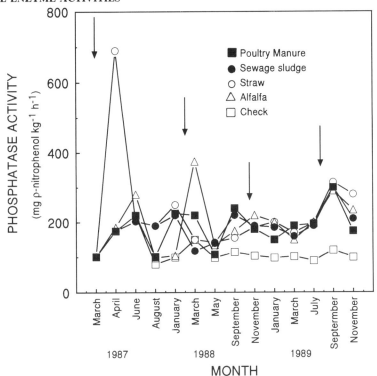

Fig. 7-4. Influence of organic amendments on soil acid phosphatase activity. LSD (0.05) = 63. Arrows indicate addition of organic amendments at 25 Mg ha^{-1} per application; dry weight basis (adapted from Martens et al., 1992).

interpretation difficult. There is relatively little field-based information, but preliminary investigations seem to suggest that once a particular management system has been established, activity of some enzymes eventually tends to stabilize. When enzyme activities were measured over a 31-mo period in soils (free of vegetation) amended with organic residues, all enzymes showed a drastic increase in activity after the first residue application (Martens et al., 1992). Even with further residue applications, activities stabilized at lower levels than the initial spike of activity within 1 yr for acid phosphatase, N-acetyl-β-glucosaminidase (EC 3.2.1.30), β-glucosidase, β-galactosidase (EC 3.2.1.23), and urease (see Fig. 7-4 as a typical response of these enzymes), and within 2 yr for amidase. In general, the stabilized level of activity after 31 mo was still significantly higher than the control.

From these results, Martens et al. (1992) hypothesized that, after the initial stimulatory effect of the new substrate, production of high levels of enzymes is inhibited by a feedback mechanism due to an adequate supply of energy. It might also be possible that there is an increase in the amount of abiontic enzymes stabilized at micro habitats that are available to carry out reactions. In such a case, microorganisms would no longer need to excrete large amounts of hydrolytic exoenzymes into soil solution to decompose substrates.

Other researchers have noted small year-to-year variations of soil enzyme activities in cropped situations (Ramírez-Martínez & McLaren, 1966; Kiss et al., 1974). However, not all enzymes exhibit stable seasonal activity, which has included alkaline phosphatase, arylsulfatase, invertase, dehydrogenase (Martens et al., 1992), and amylase (Cortez et al., 1972). Therefore, enzymes should be screened for seasonal stability before further consideration as a soil quality index.

Work by Bolton et al. (1985) allows comparison of seasonal variability of soil biological parameters in alternative or conventional cropping systems. They found enzyme activities consistently showed a highly significant treatment effect in three samplings over 1 yr, whereas microbial counts were ineffective in discriminating treatment effects. Although microbial biomass (fumigation procedure) could discriminate treatment effects (particularly N flush biomass), annual variability was lower for phosphatase and urease activities (2–7% coefficient of variability) (CV) than biomass C (33–50% CVs) and N flush biomass (11–15% CVs).

Interpretations

As can be seen from the above discussion, soil enzymes can be sensitive to perturbations and can show change over time or in comparative situations. However, it would be preferable if a soil quality measurement could be a reasonable estimate at a single point in time, without comparison to a control or reference to regionality. Given the history of the long and extensive research that went into the present day soil testing methodology, which still has many shortcomings and is very regional, this goal may never be possible. Nonetheless, there have been a few attempts to use soil enzyme activities as a basis for empirical indexes of soil biology or fertility.

Stefanie et al. (1984) proposed the biological index of fertility (BIF), which was calculated as follows:

$$BIF = (DHG + kCA)/2 \qquad [1]$$

where DH is dehydrogenase activity, CA is catalase activity, and k is a proportionality coefficient.

Beck (1984) proposed the enzyme number (EAN) index as follows:

$$EAN = 0.2(DH + CA/10 + AP/40 + PR/2 + AM/20) \qquad [2]$$

where DH is dehydrogenase activity (g TPF 10 kg^{-1} 27 h^{-1}), CA is catalase activity (% O_2 3 min^{-1}), AP is alkaline phosphatase activity (mg PNP 10 kg^{-1} 5 h^{-1}), PR is protease activity (g amino N 10 kg^{-1} 16 h^{-1}), and AM is amylase activity (% starch decomposed, 10 g^{-1} 16 h^{-1}).

Beck (1984) reported EAN values ranging from 1 to 4 for cultivated soils and 2 to 8 for pastures and forest soils.

Perucci (1992) proposed a hydrolyzing coefficient (HC) based on the ratio between the μmol of fluorescein diacetate hydrolyzed by a soil and μmol of total fluorescein diacetate added to the assay. Potential values range from 0 to 1. In this 3-yr study, HC values of 0.142 were found for the control and 0.218 to 0.367 or 0.245 to 0.442 for soils amended with 30 or 90 Mg municipal refuse ha^{-1} yr^{-1}, respectively. Using these same soil samples, Perucci (1992) generated data to calculate BIF and EAN values. The BIF showed no significant correlations with enzyme activities or biomass C (fumigation procedure), whereas EAN and HC were significatnly correlated with biomass C (r = 0.84 and 0.92, respectively). The advantage of the HC is that it is a single analysis, whereas the EAN has five enzyme analyses. Considerably more research on HC on a variety of soils and ecosystems is required before any firm conclusions can be drawn on its application as a soil fertility or quality index.

SOIL ENZYME ASSAYS

In discussing the potential of using soil enzymes to measure soil quality, one must consider the issue of methods and standard procedural protocols. Because of limited success in extracting enzymes from soils (Tabatabai, 1982), it has been necessary to study activities of soil enzymes. The assay involves adding a standard amount of soil to a known concentration of substrate and measuring the rate of disappearance of substrate or production of product. Most assays measure the product, which is preferable because measuring the change in substrate concentration is often small compared with the beginning substrate concentration. Usually an agent is used to inhibit microbial growth during the assay. Choice of inhibitor can affect enzyme activity results (Skujinš, 1978). The most widely used reagent is toluene, which can inhibit some oxidoreductases but is without effect on most other enzymes (Skujinš, 1967). Its usefulness is restricted to assays that last a few hours; longer assays may risk microbial proliferation (Tabatabai, 1982).

Because soil enzyme assays are done in vitro under controlled conditions of temperature, buffer pH and incubation time, each of which can affect enzyme activity rates, enzyme activity is operationally defined. Because of these optimal conditions and an assay mixture saturated with substrate, these assays measure potential activity, and it is not possible to determine in situ enzyme rates. Although most reports of enzyme assays are done on static samples, stirring or shaking of the assay mixture can affect the results (Ladd, 1978).

Soil sample pretreatment and storage is an important factor that affects soil enzyme activities. Both air-drying and field moist samples have been used for enzyme assays. Generally it is preferred to use field moist samples to provide a better relationship with field conditions. Air-drying usually decreases soil enzyme activities, but in some cases activity can increase (Tabatabai, 1982).

From a practical perspective, air-drying facilitates rapid processing of samples. Whether this is possible may depend on the enzyme (some enzymes are inactivated by air-drying) and on the purpose for measuring enzyme activities. If the purpose is to emphasize the activity of abiontic activity, then air-drying may be advantageous, because air-drying likely interactivates exoenzymes found in soil solution and possibly those in viable cells. This could be an advantage of soil enzyme activity measurements over other biological indexes. Air-drying would likely diminish enzyme activity stimulated by recent soil amendments and emphasize the long-term biological activity by measuring the activity of accumulated enzymes complexed in the soil matrix. Conversely, if the goal is to assess exoenzymes in soil solution or in viable cells, field moist samples would be preferable.

As one reviews the literature, considerable variation exists among enzyme procedures used by various researchers, making actual activity comparisons among sites difficult. If soil enzyme activities are to be used as soil quality indicators, a standard set of procedural and pretreatment protocols and activity units needs to be established.

Another factor to consider for enzyme activities is the potential for catalysis of substates by inorganic components of soils. It is well established that hydrogen peroxide, which is used as the substrate in catalase activity, can be decomposed by inorganic constituents. The extent and mechanisms for other substrates are less known (Skujinš, 1978), and good methodologies for assessing inorganic catalysis are lacking. An average of 11% of the total rhodanese activity was associated with inorganic catalysis (as measured in steam sterilized soils) in 33 soils (Dick & Deng, 1991). Except for those substrates with relatively low energies of activation, most of the activity for common substrates involved in hydrolytic reactions are likely to be catalyzed by enzymes in soils. Still, the potential for inorganic catalysis should be assessed when selecting enzyme activities for soil quality measurements. High rates of inorganic catalysis would overestimate the biochemical activity of a given soil.

PERSPECTIVES

Enzymes exist in soil in the biotic form associated with viable microorganisms or soil fauna, and in forms not associated with living cells (abiontic) as excreted enzymes, dead cells, or complexed with organic and mineral colloids. Clearly, enzymatic activity in soil is important for catalyzing innumerable reactions necessary for life processes of microorganisms in soils, decomposition of organic residues, cycling of nutrients, and formation of organic matter and soil structure. Because of the difficulty of separating abiontic and biotic activity, it is less clear what the functional role is of abiontic soil enzymes. However, Burns (1982) hypothesized that a spatial relationship between microorganisms and immobilized enzymes may play a role in substrate catalysis by hydrolyzing substrates that are too insoluble or large for microorganisms to use directly. If this is true, soil enzyme activity as a soil quality index may not only play a role in determining active soil biology, but also suggests soils may have optimal levels of abiontic enzymes.

Early research showed inconsistent results in relating soil enzyme activity to other soil biological properties. This was partially due to inadequate measures of soil biology (microbial counts and respiration), the specificity of most enzymes for catalyzing certain reactions, and/or abiontic soil enzyme activity. With the development of new soil biological parameters, there has been a more consistent correlation of microbial parameters with the activities of some soil enzymes. However, soil enzyme activities' primary function may not be to measure biological activity per se, but rather as an integrative indicator of a change in the biology and biochemistry of soil due to external management or environmental factors. In particular, soil enzyme measurements that are less specific and hydrolytic (e.g., protease activity and the hydrolysis of fluorescein diacetate) may be best suited for such a role. Numerous examples showed soil enzyme activities to be sensitive to changes in various land management practices including residue management, soil compaction, tillage, and crop rotation. Changes in certain enzymes often occur within months or 1 yr from the time a soil treatment is applied. Thus, the potential exists for soil enzymes to identify positive or negative effects of land management practices in reasonable time periods and do so long before there are measurable changes in soil organic matter.

A major research effort is needed to develop a relative or universal index that would be interpretable without several measurements over time or comparisons among treatments. This approach makes it unlikely that a single absolute soil enzyme activity or any other biological measurement could be used to assess soil quality, because soils naturally vary widely in biological activity. For example, a well-managed sandy soil might have optimal biological activity but still have lower biological levels than a poorly managed, heavy-textured soil. This points to the need to develop biological activity coefficients or ratios as potential soil quality indicators. In the case of soil enzymes, ratios of activity to organic C or clay content may be appropriate. A further goal of a universal index is that it should relate to plant productivity. Soil enzyme activities have not consistently correlated to crop yields, which is partially related to external input factors such as fertilizer and water, which can have a major impact on plant productivity but not on soil biological activity. This points to the need for a universal index that is a composite of key soil biological, chemical, and physical parameters.

In conclusion, the activity measurement of selected soil enzymes holds potential as a soil quality indicator because: many enzyme procedures are relatively simple, rapid, and could be done on a routine basis; they are sensitive to temporal changes in soils due to environmental and management factors; and an abiontic component can provide information on the long-term effects of management on soils. Before soil enzyme activities can be used as soil quality indices, soil sample pretreatment, assay procedures, and units of measurement must be standardized. Systematic studies across ecosystems and long-term soil management sites are also needed to identify the most appropriate soil quality enzyme assays, and to provide data for calibration and interpretation of these assays as independent soil quality indexes or as a component of a larger universal index that includes other soil properties.

REFERENCES

Alef, K., Th. Beck, L. Zelles, and D. Kleiner. 1988. A comparison of methods to estimate microbial biomass and N-mineralization in agricultural and grassland soils. Soil Biol. Biochem. 20:561-565.

Alexander, M. 1977. Microbiology of the rhizosphere. p. 423-437. *In* Introduction to soil microbiology. 2nd ed. John Wiley & Sons, New York.

Asmar, F., F. Eiland, and N.E. Nielsen. 1992. Interrelationship between extracellular enzyme activity, ATP content, total counts of bacteria and CO_2 evolution. Soil Biol. Fert. 14:288-292.

Beck, T. 1984. Methods and application of soil microbiological analysis at the Landensanstalt für Bodenkultur und Pflanzenbau (LBB) in Munich for the determination of some aspects of soil fertility. p. 13-20. *In* 5th Symp. on Soil Biology, Bucharest, Romania. February 1981. Romanian National Soc. of Soil Sci., Bucharest.

Blagoveshchenskaya, Z.K., and N.A. Danchenko. 1974. Activity of soil enzymes after prolonged application of fertilizers to a corn monoculture and crops in rotation. Sov. Soil. Sci. 5:569-575.

Bolton, H., Jr., L.F. Elliot, R.I. Papendick, and D.F. Bezdicek. 1985. Soil microbial biomass and selected soil enzyme activities: Effect of fertilization and cropping practices. Soil Biol. Biochem. 17:297-302.

Bonmati, M., H. Pujola, J. Sana, F. Soliva, M.T. Felipo, M. Garau, B. Ceccanti, and P. Nannipieri. 1985. Chemical properties, populations of nitrite oxidizers, urease and phosphate activities in sewage sludge-amended soil. Plant Soil 84:79-91.

Boyd, S.A., and M.M. Mortland. 1990. Enzyme interactions with clays and clay-organic matter complexes. *In* J.M. Bollag and G. Stotzky (ed.) Soil Biochem. 6:1-28.

Burns, R.G. 1982. Enzyme activity in soil: Location and a possible role in microbial activity. Soil Biol. Biochem. 14:423-427.

Capriel, P., T. Beck, H. Borchert, and P. Härter. 1990. Relationship between soil aliphatic fraction extracted with supercritical hexane, soil microbial biomass, and soil aggregate stability. Soil Sci. Soc. Am. J. 54:415-420.

Castellano, S.D., and R.P. Dick. 1991. Influence of cropping and sulfur fertilization on transformations of sulfur in soils. Soil Sci. Soc. Am. J. 55:283-285.

Chesters, G., O.J. Attoe, and O.N. Allen. 1957. Soil aggregation in relation to various soil constituents. Soil Sci. Soc. Am. Proc. 21:272-277.

Chunderova, A.I., and T. Zubeta. 1969. Phosphatase activity in dernopodzolic soils. Pochvovendenie 11:47-53. (Soils Fert. 33:2092).

Cortez, J., P. Lossaint, and G. Billés. 1972. Biological activity of soils in the Mediterranean ecosystems: III. Enzymatic activities. Rev. Ecol. Biol. Sol. 9:1-19.

Dick, R.P. 1992. A review: Long-term effects of agricultural systems on soil biochemical and microbial parameters. Agric. Ecosyst. Environ. 40:25-36.

Dick, R.P., and S. Deng. 1991. Multivariate factor analysis of sulfur oxidation and rhodanese activity in soils. Biogeochemistry 12:87-101.

Dick, R.P., D.D. Myrold, and E.A. Kerle. 1988a. Microbial biomass and soil enzyme activities in compacted and rehabilitated skid trail soils. Soil Sci. Soc. Am. J. 52:512-516.

Dick, R.P., P.E. Rasmussen, and E.A. Kerle. 1988b. Influence of long-term residue management on soil enzyme activity in relation to soil chemical properties of a wheat-fallow system. Biol. Fert. Soils. 6:159-164.

Dick, R.P., and M.A. Tabatabai. 1986. Hydrolysis of polyphosphates by corn roots. Plant Soil 94:247-256.

Dick, W.A., 1984. Influence of long-term tillage and crop rotation combinations on soil enzyme activities. Soil Sci. Soc. Am. J. 48:569-574.

Dick, W.A., N.G. Juma, and M.A. Tabatabai. 1983. Effects of soils on acid phosphatase and inorganic pyrophosphatase of corn roots. Soil Sci. 136:19-25.

Dick, W.A., and M.A. Tabatabai. 1992. Potential uses of soil enzymes. p. 95-127. *In* F.B. Metting, Jr. (ed.) Soil microbial ecology: Applications in agricultural and environmental management. Marcel Dekker, New York.

Doran, J.W. 1980. Soil microbial and biochemical changes associated with reduced tillage. Soil Sci. Soc. Am. J. 44:765-771.

Drobník, J. 1957. Biological transformations of organic substances in the soil. Pochvovedenie 12:62-71.

Estermann, E.F., and A.D. McLaren. 1961. Contribution of rhizoplane organisms to the total capacity of plants to utilize organic nutrients. Plant Soil 15:243-260.

Frankenberger, W.T., Jr., and W.A. Dick. 1983. Relationships between enzyme activities and microbial growth and activity indices in soil. Soil Sci. Soc. Am. J. 47:945-951.

Frankenberger, W.T., Jr., J.B. Johanson, and C.O. Nelson. 1983. Urease activity in sewage sludge-amended soils. Soil Biol. Biochem. 15:543-549.

Galstyan, A.Sh. 1960. Enzyme activities in solchaks. Dokl. Akad. Nauk Arm. SSR 30:61-64.

Guilbaut, G.G., and D.N. Kramer. 1964. Fluorometric determination of lipase, acrylase, alpha- and gama-chymotrypsin and inhibitors of these enzymes. Anal. Chem. 36:409-412.

Gupta, V.V.S.R., and J.J. Germida. 1988. Distribution of microbial biomass and its activity in different soil aggregate size classes as affected by cultivation. Soil Biol. Biochem. 20:777-786.

Habán, L. 1967. Effect of ploughing depth and cultivated crops on the soil microflora and enzyme activity of the soil. Ved. Pr. Vysk. Ustava Rastl. Vyroby Piestanoch 5:159-169.

Howard, P.J.A. 1972. Problems in the estimation of biological activity in soil. Oikos 23:235-240.

Khan, S.U. 1970. Enzymatic activity in a gray wooded soil as influenced by cropping systems and fertilizers. Soil Biol. Biochem. 2:137-139.

Kiss, I. 1957. The invertase activity of earthworm casts and soils from ant-hills. Agrokem. Talajtan 6:65-58.

Kiss, I., M. Drăgan-Bularda, and D. Radulescu. 1975. Biological significance of enzymes in soil. Adv. Agron. 27:25-87.

Kiss, S., G. Stefonic, and M. Drăgan-Bularda. 1974. Soil enzymology in Romaina. Part 2. Contrib. Bot. Univ. "Babes Bolyai" Cluj Napoca Gradina Bot., 197-207.

Klein, T.M., and J.S. Koths. 1980. Urease, protease, and phosphatase in soil continuously cropped to corn by conventional or no-tillage methods. Soil Biol. Biochem. 12:293-294.

Koepf, H. 1954a. Investigations on the biological activity in soil: I. Respiration curves of the soil and enzyme activity under the influence of fertilizing and plant growth. Z. Acker Pflanzenbau 98:289-312.

Koepf, H. 1954b. Experimental study of soil evaluation by biochemical reactions: II. Enzyme reactions and CO_2 evolution in different soils. Z. Pflanzenernahr. Dung. Bodenkd. 67:262-270.

Kozlov, K.A. 1965. The role of soil fauna in the enrichment of soil with enzymes. Pedobiologia 5:140-145.

Ladd, J.N. 1978. Origin and range of enzymes in soil. p. 51-96. In Origin and range of enzymes in soil. p. 51-96. In R.G. Burns (ed.) Soil enzymes. Academic Press, New York.

Laugesen, K. 1972. Urease activity in Danish soils. Dan. J. Plant Soil. Sci. 76:221-229.

Laugesen, K., and J.P. Mikkelsen. 1973. Phosphatase activity in Danish soils. Dan. J. Plant Soil Sci. 77:252-257.

Lundgren, B. 1981. Fluorescein diacetate as a stain or metabolically active bacteria in soil. Oikos 36:17-22.

Martens, D.A., J.B. Johanson, and W.T. Frankenberger, Jr. 1992. Production and persistence of soil enzymes with repeated additions of organic residues. Soil Sci. 153:53-61.

Medzon, E.L., and M.L. Brady. 1969. Direct measurement of acetylesterase in living protist cells. J. Bacteriol. 97:402-415.

Nannipieri, P., R.L. Johnson, and E.A. Paul. 1978. Criteria for measurement of microbial-growth and activity in soil. Soil Biol. Biochem. 10:223-227.

Nannipieri, P., L. Muccini, and C. Ciardi. 1983. Microbial biomass and enzyme activities: Production and persistence. Soil Biol. Biochem. 15:679-685.

Park, S.C., T.J. Smith, and M.S. Bisesi. 1990. Hydrolysis of bis(4-nitrophenyl)phosphate by the earth worm Lumbricus terrestris. Soil Biol. Biochem. 22:729-730.

Park, S.C., T.J. Smith, and M.S. Bisesi. 1992. Activities of phosphomonoesterase from Lumbricus terrestris. Soil Biol. Biochem. 24:873-876.

Paul, E.A., and A.D. McLaren. 1975. Biochemistry of the soil subsystem. p. 1-36. In E.A. Paul and A.D. McLaren (ed.) Soil biochemistry. Vol. 3. Marcel Dekker, New York.

Perucci, P. 1992. Enzyme activity and microbial biomass in a field soil amended with municipal refuse. Biol. Fert. Soils. 14:54-60.

Quastel, J.H. 1946. Soil metabolism. The Royal Inst. of Chem. of Great Britain and Ireland, London.

Ramírez-Martínez, J.R., and A.D. McLaren. 1966. Some factors influencing the determination of phosphatase activity in native soils and in soils sterilized by irradiation. Enzymologia 31:23-38.

Reganold, J.P. 1988. Comparison of soil properties as influenced by organic and conventional farming systems. Am. J. Altern. Agric. 3:144-155.

Rogers, H.T. 1942. The availability of certain forms of organic phosphorus compounds by soil catalysts. Iowa State Coll. J. Sci. 17:108-110.

Rogers, H.T., R.W. Pearson, and W.H. Pierre. 1942. The source and phosphatase activity of exoenzyme systems of corn and tomato roots. Soil Sci. 54:353-366.

Ross, D.J., and A. Cairns. 1982. Effect of earthworms and ryegrass on respiratory and enzyme activities of soil. Soil Biol. Biochem. 14:583-587.

Rotman, B., and B.W. Papermaster. 1966. Membrane properties of living mammalian cells as studied by enzymatic hydrolysis of fluorogenic ester. Proc. Natl. Acad. Sci. U.S.. 55:134-141.

Schnürer, J., and T. Rosswall. 1982. Fluorescein diacetate hydrolysis as a measure of total microbial activity in soil and litter. Appl. Environ. Microbiol. 43:1256-1261.

Sharpley, A.N., and J.K. Syers. 1977. Potential role of earthworm casts for the phosphorus enrichment of runoff waters. Soil Biol. Biochem. 8:341-346.

Skujinš, J. 1976. Enzymes in soil. p. 371-414. In A.D. McLaren and G.H. Peterson (ed.) Soil biochemistry. Marcel Dekker, New York.

Skujinš, J. 1976. Extracellular enzymes in soil. CRC Crit. Rev. Microbiol. 4:383-421.

Skujinš, J. 1978. History of abiontic soil enzyme research. p. 1-49. In R.G. Burns (ed.) Soil enzymes. Academic Press, New York.

Söderström, B.E. 1977. Vital staining of fungi in pure cultures and in soil with fluorescein diacetate. Soil Biol. Biochem. 9:59-63.

Speir, T.W. 1977. Studies on a climosequence of soils in tussock grasslands: XI. Urease, phosphatase and sulfatase activities of topsoils and their relationships with other properties including plant available sulfur. N.Z. J. Sci. 20:159-166.

Speir, T.W., and D.J. Ross. 1976. Studies on a climosequence of soils in tussock grasslands: IX. Influence of age of Chinochloa rigida on enzyme activities. N.Z. J. Sci. 19:389-396.

Speir, T.W., R. Lee, E.A. Pansier, and A. Cairns. 1980. A comparison of sulphatase, urease and protease activities in planted and fallow soils. Soil Biol. Biochem. 12:281-291.

Stefanic, F., G. Eliade, and I. Chirnogeanu. 1984. Researches concerning a biological index of soil fertility. p. 35-45. In 5th Symp. on Soil Biology. Bucharest, Romaina. February 1981. Romanian National Soc. of Soil Sci., Bucharest.

Syers, J.K., A.N. Sharpley, and D.R. Keeney. 1979. Cycling of nitrogen by surface-casting earthworms in a pasture ecosystem. Soil Biol. Biochem. 11:181-185.

Tabatabai, M.A. 1982. Soil enzymes. p. 903-945. In A.L. Page et al. (ed.) Methods of soil analysis. Part 2. 2nd ed. Agron. Monog. 9. ASA and SSSA, Madison, WI.

Tiwari, M.B., B.K. Tiwari, and R.R. Mishra. 1989. Enzyme activity and carbon dioxide evolution from upland and wetland rice soils under three agricultural practices in hilly regions. Biol. Fert. Soils. 7:359-364.

Verstraete, W., and J.P. Voets. 1977. Soil microbial and biochemical characteristics in relation to soil management and fertility. Soil Biol. Biochem. 9:253-258.

Werner, W., H.W. Scherer, and H.W. Olfs. 1988. Influence of long-term application of sewage sludge and compost from garbage with sewage sludge on soil fertility criteria. J. Agron. Crop Sci. 160:173-179.

Yaroschevich, I.V. 1966. Effect of fifty years' application of fertilizers in a rotation on the biological activity of a chernozem. Agrokhimiya 6:14-19.

8 Assessment and Significance of Biologically Active Soil Organic Nitrogen[1]

John M. Duxbury and Sipho V. Nkambule

Cornell University
Ithaca, New York

Interest in active soil organic N arises because of its relationship to the capability of soils to supply N for crop growth and its possible use as an index of soil quality. Active soil organic N is also linked to agricultural sustainability and environmental quality (Fig. 8-1), because losses of N from soil can lead to pollution of water and the atmosphere. Consequently, an important goal of sustainable agriculture is to minimize undesirable environmental side effects of N through management of active soil organic N in coordination with added N sources. This will undoubtedly require coupling knowledge of controls on mineralization of soil organic N with appropriate cropping systems and soil management practices. Key objectives to strive for will be synchrony between N supply from soil and demand for N in soil–crop systems, and conservation of N mineralized outside normal crop growth periods.

Active soil organic N can be defined as a single pool of N, or it can be broken down into several components based on current concepts of soil organic matter dynamics (Duxbury et al., 1991). A simplified diagram of the soil N cycle relevant to active soil organic N is shown in Fig. 8-2. It shows a biomass pool and three soil organic N pools that differ in their susceptibility to biological decomposition. The active pool consists of soil organic N compounds that are free within the soil matrix, whereas the stabilized and passive pools contain organic matter that is inaccessible to soil organisms and enzymes because of interactions with the soil matrix. As implied by its name, the passive N pool is essentially inert to biological decomposition, perhaps becuse of molecular scale interactions between organic matter and soil mineral components; these are thought to be the main driving force in the formation of small aggregates (< 20 μm in size), which are very stable (Tisdall & Oades, 1982). In contrast, the stabilized N pool is protected by physical incorporation into medium- to large-sized aggregates, up to several

[1] USDSA/CSRS is acknowledged for financial support of Cornell's Agricultural Ecosystems Program, which carried out some of the research reported in this chapter.

ACTIVE SOIL NITROGEN

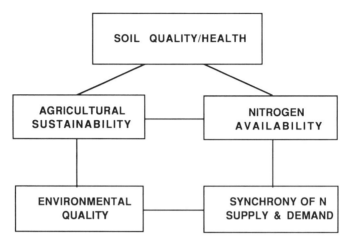

Fig. 8–1. Conceptual relationships between soil properties and processes that include active soil N.

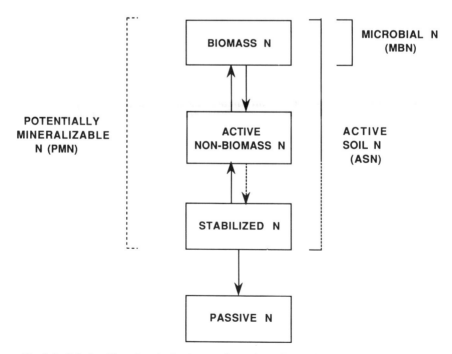

Fig. 8–2. Relationships of methods of measuring active soil organic N to pools of soil organic N used in modeling organic matter dynamics.

millimeters in size. Nitrogen in the stabilized pool can be converted into active N by disruption of these aggregates, which occurs to varying degrees with wet/dry or freeze/thaw cycles and with tillage.

Although the discrete pool approach provides a conceptual framework within which to evaluate methods of measuring the biological availability of soil organic N and is especially useful for mathematical modeling of soil organic matter dynamics, it should be recognized that it is an approximation. In reality, a continuum of stability of soil organic matter and soil organic N probably exists, and this complicates measurement of active soil organic N.

The objectives of this chapter are to:

1. Evaluate methods of measuring biomass N and active organic N pools
2. Assess the utility of biomass N and active soil organic N measurements as indices of soil quality and as predictors of soil N availability to crops
3. Describe and discuss relationships between the various measurements

The biomass N pool is described as microbial biomass N (MBN), using the term *microbial* in a broad way because of the focus on $CHCl_3$ fumigation methods of measurement (Jenkinson, 1988). Because $CHCl_3$ is a general fumigant, the MBN measurement will potentially include all organisms that are in a soil sample. Given that coarse sieving of soils, usually through a 2- or 4-mm sieve, is the most common soil preparation practice, microflora (bacteria and fungi), microfauna (especially protozoa and nematodes), and mesofauna (e.g., Collembola and mites) will be included, but larger soil animals (e.g., earthworms and beetles) will be excluded. However, micro- and mesofauna normally account for <3% of the microbial N (Verhoef & Brussaard, 1990) so that the MBN measurement is essentially that of microflora N. The other N parameters considered are potentially mineralizable N (PMN) as developed by Stanford and Smith (1972) and active soil organic N (ASN) determined by an isotope dilution method (Duxbury et al., 1991).

Before evaluating these methods, it is worth considering what they measure in terms of the N pools identified in Fig. 8-2. The measurement of soil microbial N presumably measures only this pool as described above. In the isotope dilution method, *active soil organic nitrogen* is defined as that N which participates in biologically mediated N cycling within soils, and it relies on microbial activity to mix added ^{15}N with the ASN pool until isotopic equilibrium is obtained. Consequently, this method measures active nonbiomass N and the microbial biomass involved in the mixing process. The PMN determination is based on measurement of N mineralized in laboratory incubation and mathematical fitting of pool size(s) and rate constant(s) to these data, usually using first order kinetics. Although the PMN measurement does not directly measure any of the N pools shown in Fig. 8-2, it likely includes the bulk of the active–nonbiomass N and some fraction of the microbial N in the soil at the time of sampling, because there is evidence that the microbial biomass can decline under the incubation conditions.

Sample handling and processing procedures also need to be evaluated, because these have the potential to convert some of the stabilized N into the active N pool. Should this occur, both PMN and ASN measurements would include some unknown portion of the stabilized N pool.

EVALUATION OF METHODOLOGIES

Microbial Biomass Nitrogen

Methods of selectively measuring the bacterial and fungal biomass in soils include direct observation and probably substrate-induced respiration (Schmidt & Paul, 1982; Parkinson & Paul, 1982). Larger organisms are usually measured directly by a variety of procedures (see Page et al., 1982). Measurements of ATP level (Parkinson & Paul, 1982) and the $CHCl_3$ fumigation–incubation or $CHCl_3$ fumigation–extraction method (Jenkinson, 1988) span a range of organisms as described in the preceding section. Of all these methods, only the fumigation-based procedures can easily measure microbial N (Shen et al., 1984; Brookes et al., 1985). Although the fumigation procedures are widely used, questions about their accuracy, reproducibility, and utility still remain.

A central issue with the fumigation methods is how to choose the conversion factor (k_N) used in calculating mineralized or extracted N to MBN. A limitation of the fumigation–incubation method is that the quantity of N mineralized in fumigated soil depends somewhat on the C/N ratio of the microbial substrate, which, for various reasons, can be wide enough in some soils to interfere with the measurement (Jenkinson, 1988), and is, of course, unknown when the measurement is made. Two approaches have been taken to overcome this limitation; Voroney and Paul (1984) developed an equation that uses the ratio of C mineralized to N mineralized in calculating k_N, while Brookes et al. (1985) developed a fumigation–extraction method that avoids incubation but relies on calibration to the fumigation–incubation procedure for determination of a k_N value. At present, there is no generally agreed on value or procedure for determining k_N. Published values for the fumigation–incubation method range from 0.2 to 0.68 (Voroney & Paul, 1984; Shen et al., 1984; Azam et al., 1988). A single value (0.68 for a 5-d fumigation and 0.54 for a 1-d fumigation) is recommended for the fumigation–extraction method (Brookes et al., 1985). A weakness of the fumigation-extraction method is that the k_N is not determined independently of the fumigation–incubation method. An appropriate independent determination of this conversion factor would estimate the fraction of microbial biomass that is extracted by a salt solution immediately after fumigation. The effects of soil properties such as pH, texture, and mineralogy on the extractability of organic N compounds should also be systematically investigated.

A second issue that can be raised is that no attempt is made to standardize the chemical or biological conditions of the fumigation-based methods. Should fumigation treatment and subsequent mineralization and/or extraction be carried out under pH buffered conditions? Should inorganic nutrients be added to incubated samples? Should standardized microbial populations be used to mineralize substrates released by fumigation? Not all of these options may be desirable, but it is surprising that none of them appear to have been investigated, especially pH buffered extraction with the fumigation–extraction method.

A widespread deficiency with MBN data is a general lack of analysis of uncertainty associated with reported values. At best, the analytical variability associated with replicate measurements of MBN on the same sample is presented. However, in both fumigation procedures a conversion factor (k_N) is used to transform a measured N flush to a MBN value. Uncertainty associated with the conversion factor needs to be coupled with analytical variability to derive a combined estimate of uncertainty in the reported data. For example, we calculated the value of k_N for the fumigation–incubation procedure using the equation of Shen et al. (1984) and k_C values (the conversion factor used in calculating microbial biomass C) of four soils from Germany (Anderson & Domsch, 1978) and one soil from England (Jenkinson, 1976). Although an average k_N value of 0.65 can be derived, the 95% confidence interval for k_N is 0.56 to 0.74. If the N flush values for two soils were 20 and 25 mg N kg^{-1}, calculated MBN values using the average k_N value would be 31 and 38 mg N kg^{-1}, whereas the actual range with 95% confidence would be 27 to 36 and 34 to 45 mg N kg^{-1}. One could argue that use of a single conversion factor is justified on the basis that the uncertainty in its measurement is due to analytical variability in a multistep process; however, it is more likely that it is due to real differences between soils. Whether or not one agrees with this position, it is clear that evaluation of uncertainty in MBN measurements is needed if we are to properly interpret MBN data.

A final issue to consider, in a general sense, is the utility of MBN data derived from fumigation methods. The fumigation procedure presumably kills both active and inactive microorganisms so that the total microbial biomass N is measured—or is it? McGill et al. (1986) suggest that much of the soil microbial bioamss may be dormant or even nonviable. If one attempts to study the temporal dynamics of MBN, measuring changes in the size or activity of the active microbial biomass fraction rather than the total would be more appropriate. If one is interested in overall capacity or storage-related factors, then measuring the total microbial biomass pool would be appropriate. One example of the importance of making this distinction is that the soil fauna are estimated to be responsible for 30% of the N mineralized by microbial biomass in both natural and agricultural ecosystems, even though they contain <3% of the total MBN pool (Verhoef & Brussaard, 1990).

Potentially Mineralizable Nitrogen

Since this method was introduced by Stanford and Smith (1972), most effort has been directed at improving mathematical estimation of the PMN

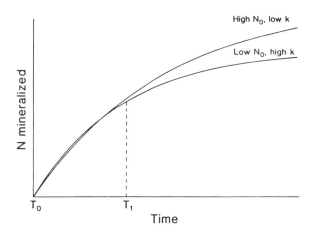

Fig. 8-3. Idealized mineralization curves showing difficulty in selecting between high N_o, low k and low N_o, high k scenarios for a short incubation time, T_1 (from Paustian & Bonde, 1987).

pool and associated kinetic parameters. The calculation process involves fitting a kinetic model to the observed mineralization pattern by simultaneously varying the size of the PMN pool and a rate constant. Stanford and Smith assumed that mineralization was a first-order kinetic process and derived a single PMN pool and a rate constant. Subsequent investigators have refined the mathematical methodology (Talpaz et al., 1981; Juma et al., 1984; Bonde & Lindberg, 1988; Sierra, 1990) and have used a double pool first-order model (Deans et al., 1986; Cabrera & Kissel, 1988a). Sometimes mineralization is linear over time (zero-order kinetics) (Tabatabai & Al-Khafaji, 1980; Addiscott, 1983; Houot et al., 1989), which prevents parameter estimation except by using a N model, such as illustrated by Houot et al. (1989).

In their original work, Stanford and Smith (1972) incubated soils for 32 wk. During this time, about 80% of the estimated PMN pool was mineralized in the 39 soils that they used. Potential problems in estimating PMN and rate constant(s) from shorter incubation periods often used by other investigators have been discussed by Paustian and Bonde (1987). As illustrated in Fig. 8-3, these include difficulty in choosing between high PMN–low k and low PMN–high k scenarios. The observation that published rate constants are generally higher and more variable among soils as incubation time is shortened (Fig. 8–4) is evidence that this difficulty exists. However, there appears to be compensatory interdependence of PMN and rate constant values, even with long-term incubations, i.e., where PMN is relatively high among replicate measurements the rate constant tends to be low and vice versa (for example, see Table 3 in Cabrera & Kissel, 1988a). This result is a consequence of the mathematical methodology, and it throws into question any independent use of these parameters. Moreover, in a study with 34 soils, Sierra (1990) found that the relationship between N mineralization rate ($\Delta N/\Delta t$) and cumulative N mineralized at time t was curvilinear rather than linear as required by a single pool exponential model. The relationship was

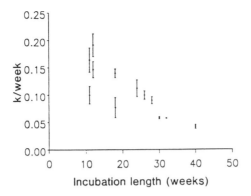

Fig. 8-4. Influence of incubation time on estimated first-order N mineralization rate constant (*k*); values show mean and standard errors (from Paustian & Bonde, 1987).

approximated by multiple pool exponential models, suggesting that the concept of a continuum of stability of soil organic N is indeed correct.

As one might expect from the well-known stimulating effects of aggregate disruption on mineralization of soil organic matter and organic N (e.g., Birch, 1958; Agarawal et al., 1971; Craswell & Waring, 1972), sample handling procedures such as sieving and air drying soils can alter the size of the measured PMN pool and the kinetics of mineralization. Cabrera and Kissel (1988a) showed that N mineralization patterns were much different when soils were air-dried and sieved (2 mm) compared with 5 cm diam. cores that were not dried. Beauchamp et al. (1986) found that air drying increased N mineralization compared with soils that were sieved field moist or after 5 d of freezing. The effect of air drying was much greater than that of freezing. Unfortunately, the effect of sieving soil without drying has been studied little; Rice et al. (1987) found enhanced mineralization in only one of three soils when soil cores were subjected to simulated tillage by sieving to <8 mm.

The dramatic effect of air drying on N mineralization is illustrated in Fig. 8-5, which shows representative results obtained by Cabrera and Kissel (1988a). It can be seen that there is an initial burst of N mineralization in sieved soils but not in soil cores. Moreover, after the initial flush, N mineralization continues at a greater rate in sieved soils than in soil cores. Some authors (e.g., Dalal & Mayer, 1987), including Stanford and Smith (1972), have assumed that sample handling effects are limited to the initial phase of the incubation and that *true* PMN values can be obtained by simply using N mineralization data obtained after the initial burst, i.e., the most slowly mineralized N pool of a two-pool kinetic model. The work of Cabrera and Kissel (1988a) shows that this assumption is not valid. It also is clear that the use of air-dried soils would lead to higher PMN values than would be found under most field conditions. Fortunately, the tendency in recent PMN studies is to avoid air drying soils.

Another question that needs to be considered is whether long-term laboratory incubations lead to results that are applicable in the real world. To answer this question, one needs to know what factors control minerali-

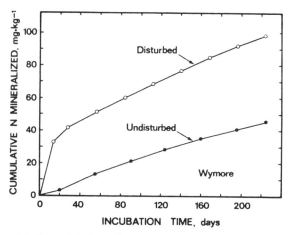

Fig. 8-5. Effect of drying and sieving soils on N mineralization in a Wymore soil (mesic Aquic Argiudoll) (from Cabrera & Kissel, 1988a).

zation of soil organic N under field conditions. Stanford and Smith assumed that temperature and moisture were the primary controls on mineralization of N from the PMN pool and established relationships between these parameters and the mineralization rate constant, k (Stanford et al., 1973; Stanford & Epstein, 1974). Nitrogen mineralization in the field could then be calculated by using field temperature and moisture conditions to modify k. This approach assumes a discrete PMN pool of known initial size. The concept of a discrete PMN pool is, however, incorrect because the availability of organic N substrate and hence N mineralization partially depends on release of stablized N to the active N pool. In fact, the availability of organic N substrate can be considered to be the primary control of N mineralization in the field. Substrate availability is influenced by soil temperature (freeze–thaw events), soil moisture (dry–wet cycles), and by tillage. These processes and their effects, however, are not predictable and/or not represented by simple field temperature and moisture measurements or by constant environment laboratory incubations.

A second deficiency of a long-term laboratory incubation is effects on the microbial population. Bonde et al. (1988) found that the microbial biomass declined substantially during a 25-wk incubation, and between 55 and 89% of the N mineralized could be directly attributed to this decline. Similarly, Carter and Rennie (1982) found that 58% of the N mineralized in a 7-wk incubation was explained by the decrease in MBN. Others (Juma & Paul, 1984; Nkambule, 1993) have found only small declines in microbial biomass during similar length incubations. More research is needed to establish the circumstances that cause large declines in microbial biomass since this would not normally occur during crop growth in the field. This is not to say that changes in the size of the microbial biomass do not occur in the field; rather, it is again the question of whether the laboratory incubation can adequately represent what happens in the field.

Active Soil Nitrogen by Isotope Dilution

The ASN method of Duxbury et al. (1991) is based on earlier work of Jansson (1958). It involves insertion of a known quantity of ^{15}N enriched NH_4^+ into the soil N cycle and stimulation of microbial activity so that the added ^{15}N will mix with the active N pool until isotopic equilibrium is reached. At this point, the ^{15}N enrichment of the NH_4^+ pool is measured, and the size of the ASN pool (x) is calculated using the following isotope dilution equation:

$$N_e \text{ (atom\% } ^{15}N) = \frac{x \, (0.366) \, + \, y \, (^{15}N \text{ atom\% of added } NH_4^+)}{x + y}$$

where N_e is the measured ^{15}N content of the NH_4^+ pool at assumed equilibrium and y is the amount of added $^{15}NH_4$.

Because the measured size of the ASN pool increases somewhat with incubation time, the term ASN is operationally defined by the method and different investigators may select different incubation times, depending on their particular objectives. For example, ASN values measured in a silt loam soil after 20 d of incubation at 30 °C were 65 to 75% of the values obtained after 70 d of incubation (J.R. Fruci & J.M. Duxbury, 1992, unpublished data). Follett and Schimel (1989) also showed the size of an active N pool measured with ^{15}N can increase with incubation time. One explanation for this result could be that the isotopic mixing process is not complete, and indeed this will always be the case unless the ASN is truly a discrete pool. However, other work (F.J.S.A. Costa and J.M. Duxbury, 1992, unpublished data) indicates that mixing of the added $^{15}NH_4^+$ with the microbial pool is completed within 20 d so that the increasing value of ASN may be attributed to the inclusion of more slowly available or less active N in the measurement. This result is consistent with the concept of a continuum of stability for soil organic N rather than discrete pools of defined stability.

An important feature of the ASN method is that stimulation of microbially mediated isotopic mixing by adding carbonaceous substrates does not interfere with the calculation of the size of the ASN pool. Stimulation of microbial mixing was achieved in two ways; in one case sufficient glucose was included to immobilize the added $^{15}N_4^+$, and in a second case this same treatment was followed by a 24-h fumigation with $CHCl_3$ after 3 d of incubation (Duxbury et al., 1991). The rationale behind the fumigation treatment was to introduce a second mixing stage involving a diverse group of substrates (dead microorganisms) after initial metabolism of the added glucose. The ASN pool size measured with the fumigation procedure was always found to be higher than that measured without fumigation. A probable explanation for this result is that fumigation introduces into the measurement microorganisms not responding to the glucose addition, that is it includes inactive microorganisms. In 10 soils from New York, ASN values for the method utilizing fumigation ranged from 113 to 283 mg N kg^{-1} soil.

The two ASN procedures were significantly correlated ($r^2 = 0.79$) and the value obtained with fumigation was 1.84 times that obtained without fumigation (Nkambule, 1993).

The ASN measurement is sensitive to the same sample handling factors that affect the PMN measurement. Thus, air-drying and rewetting of a silt loam soil from a long-term pasture gave an ASN value that was 75% greater than that obtained for the field moist soil, but sieving the soil to 1 mm had no effect on the result (J.R. Fruci & J.M. Duxbury, 1992, unpublished data).

SOIL QUALITY

The microbial biomass is both an agent of transformation of residues and soil organic matter and a sink or source of nutrients as its size varies. A consistently high level of microbial biomass C or N would be interpreted as being beneficial in terms of soil quality, whereas a consistently low level of microbial biomass would indicate poor soil quality. As illustrated by Fig. 8–6, which is modified from Campbell et al. (1991), soil MBN levels generally increase as plant cover becomes more continuous. In this example from a wheat (*Triticum aestivum* L.)-based cropping system, inclusion of a green manure or switching to a rotation that has a pasture phase led to the highest MBN values. This result is clearly consistent with expectations that these types of modifications to a small grain production system would increase soil quality.

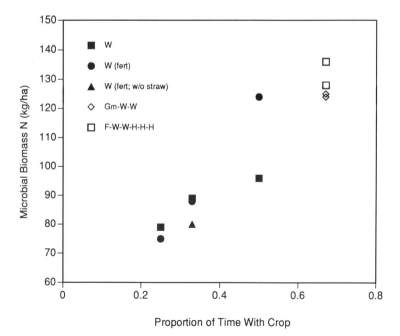

Fig. 8–6. Effect of frequency of crop cover on soil microbial biomass N; W = wheat, *Gm* = green manure, F = fallow, and H = hay (modified from Campbell et al., 1991).

Table 8-1. Effect of crop rotation on soil microbial biomass N levels and dynamics (data from McGill et al., 1986).

Rotation	Crop	Microbial biomass		
		Mean§	Δ(NPK)¶	Δ(cont)¶
			mg kg^{-1}	
WOBPP†	Pasture‡	57a	−18	−18
	Barley	20b	+14	+13
WF†	Wheat	23b	±6	±6
	Fallow	12c	±7	±5

† WOBPP and WF are wheat–oat–barley–pasture–pasture and wheat–fallow rotations, respectively.
‡ Second year.
§ Values followed by different letters are significantly different.
¶ Δ represents change from the initial value over the measurement period for fertilized (N–P–K) and unfertilized plots (cont).

One difficulty in using microbial biomass as an indicator of soil quality, however, is that its size is a function of organic substrate availability and can be temporally elevated, even in degraded soils. Nevertheless, it may still be possible to use MBN as an indicator of soil quality, provided that its temporal dynamics are considered. The data in Table 8-1, showing the effect of crop rotation on MBN levels and dynamics, can be used to illustrate how this might be done. Although MBN levels varied over the growing season in all systems, mean MBN levels were higher for the second year of pasture (57 mg N kg^{-1} soil) in the wheat–oat (*Avena sativa* L.)–barley (*Hordeum vulgare* L.)–pasture–pasture rotation compared with either phase of the wheat–fallow rotation (23 and 12 mg N kg^{-1} soil, respectively). However, mean MBN levels were the same in the small grain phase of either rotation, so comparisons made at this stage only would lead to the conclusion that the two rotations were equivalent in terms of soil quality. Furthermore, there is sufficient change in MBN values during the growing season (last two columns of the table) that a single measurement would be insufficient to characterize either system.

Further complicating the use of MBN or MBC as soil quality indicators is the observation that the temporal dynamics of microbial biomass appear to be as much related to soil moisture conditions as to anything else (e.g., McGill et al., 1986; Doran, 1987; Wardle & Parkinson, 1990). And finally, there is a need to assess microbial biomass levels, either N or C, on a time scale of days rather than the week to month time scale used by most investigators to better understand the meaning of such measurements. Overall, it is clear that a considerable amount of work may be needed to properly use MBN measurements as an indicator of soil quality.

Both PMN and ASN measurements are conceptually attractive as potential indicators of soil quality, because changes in these N pools will likely be detectable before changes in total soil N or organic matter levels can be measured. Thus, measurement of active soil N could provide a sensitive method for determining whether soil quality is improving or degrading. An

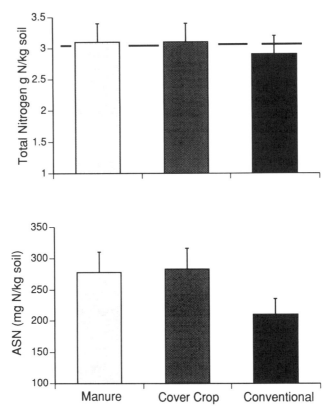

Fig. 8-7. Total and active soil N levels after 10 yr of the Rodale farming systems trial; bars show standard errors and the dashed line in the upper figure represents the normalized soil N level at the beginning of the experiment (Friedman, 1993).

example of this is given in Fig. 8–7, which shows results after 10 yr of a farming systems experiment at the Rodale Institute in Pennsylvania. This experiment compares a conventional corn (*Zea mays* L.)–soybean [*Glycine max* (L.) Merr.] system with two systems that include corn and soybean in more complex rotations and use either manure or cover crops (Liebhardt et al., 1989). The systems using manure or cover crops are considered to be more sustainable than the conventional system. Differences in ASN levels between the two sustainable systems and the conventional system were more readily discernable than were differences in total N contents, although both were significant ($p < 0.05$). Similarly, Doran (1987) found significantly higher PMN levels in no-till surface soils (0–7.5 cm) compared with conventionally tilled soils of corn tillage experiments in several midwestern U.S. states (Fig. 8–8), when differences in total N were not always found.

It is probable that PMN and ASN pools are not as temporally variable as MBN so that they would be more suitable indicators of soil quality than MBN. However, there is little data available on the dynamics of either PMN

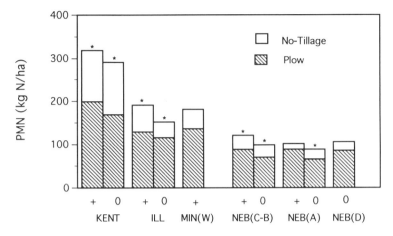

Fig. 8–8. Effect of tillage on potentially mineralizable N levels in the 0 to 7.5 cm soil depth with (+) and without (0) fertilizer addition; * indicates significant difference between tillage systems at $p < 0.05$ (from Doran, 1987).

or ASN within an annual cycle (see discussion in following section), and more research is needed to establish whether or not this is the case.

NITROGEN AVAILABILITY

One of the major limitations in the field of soil fertility is the inability to accurately predict how much N soils will provide to crops. This task has become more critical as evidence mounts that improper use of N in agriculture can lead to a variety of environmental pollution problems as well as wasting resources. Ideally, we would like to predict both the amount and timing of N release from soils. Since mineralization of soil organic N is a biological process, some appropriate biological measurement could be a good predictor of this process.

Potentially Mineralizable Nitrogen. Predicting N mineralization under field conditions and developing a simple N soil test were the ultimate objectives of the work on PMN by Stanford and his associates. The PMN method has not been used much, presumably because its measurement by even the short-term incubation method of Stanford et al. (1974) is impractical as a soil test and no widely applicable chemical surrogate for its measurement has been found (Stanford & Smith, 1978). Curiously, little effort appears to have been made to demonstrate that the size of the PMN pool changes as N is mineralized in soils, although this is perhaps the most basic consequence of the concept. In a study of four cropping systems, Bonde and Rosswall (1987) found inconsistent temporal patterns in PMN (actually N mineralized in 13 wk, but called PMN by the authors) over the growing season and within season variability was as great as differences between the systems.

The limited evaluation of PMN methodology under field conditions has led to mixed results. Several studies gave promising results (Smith et al., 1977; Stanford et al., 1977; Herlihy, 1979; Richter et al., 1980; Marion et al., 1981), but others show a substantial overprediction in N mineralization (Verstraete & Voets, 1976; Cabrera & Kissel, 1988b). Additionally, calculations of N mineralization under field conditions by coupling measured soil temperature and moisture values with PMN values chosen on the basis of total soil N content (PMN ranges from 5 to 40% of total soil N) and Stanford and Smith's rate constant (the lowest of those reported) generally give implausibly high results. The principal cause for overprediction of N mineralization by the PMN method is undoubtedly the use of air-dry soils (see earlier discussion), which creates a pool of mineralizable N that does not normally exist under field conditions (Cabrera & Kissel, 1988b) and elevates rate constant values (Beauchamp et al., 1986). Inadequate correction for moisture variability (Cabrera & Kissel, 1988b) and temperature by moisture interactions (Cassman & Munns, 1980) are probably secondary factors.

Microbial Biomass Nitrogen and Active Soil Nitrogen. Microbial biomass N could be related to the N mineralization capacity of soils, because microbes are important mineralization agents or because changes in MBN pool size reflect N mineralization or immobilization. The active soil N pool, which includes the microbial biomass, could also be related to N mineralization for similar reasons. From a management perspective, it would be useful to understand whether any relationships between these measurements and N mineralization are due predominantly to decreases in pool size or because pool size is proportional to the flux of N, i.e., it is the engine driving the mineralization process.

Changes in MBN pool size during crop growth have been determined in a number of studies, but no consistent pattern has emerged. Conflicting patterns in MBN dynamics have sometimes been observed; experiments with barley by McGill et al. (1986) and Burton and McGill (1991) on two different soils gave opposite results (Fig. 8-9). With the Chernozem soil (Mollisol), MBN declined between June and November, a period that extends beyond crop harvest, whereas it increased during this same time period with the Gray Luvisol (Burton & McGill, 1991). Thus, the microbial biomass appears to be a source of N in one experiment but a sink for N in a similar experiment.

An example of changes in soil MBN and also MBC pool sizes over a complete annual cycle is shown in Fig. 8-10. This study, carried out by Patra et al. (1990), utilized a wheat field and a pasture from the classical, long-term experiments at Rothamsted, England. Fluctuations of about 50 kg N ha^{-1} were found for the wheat field, but there does not appear to be any readily interpretable pattern to the data. A drop in soil MBN of more than 100 kg N ha^{-1} was observed in the grass field in September, but this is likely an abberant data point because no similar change was observed in soil MBC. If this point is disregarded, it would appear that the soil MBN pool size in

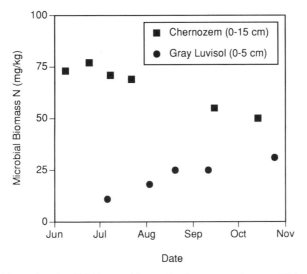

Fig. 8-9. Change in microbial biomass N over time in two experiments with barley (data from McGill et al., 1986, and Burton & McGill, 1991).

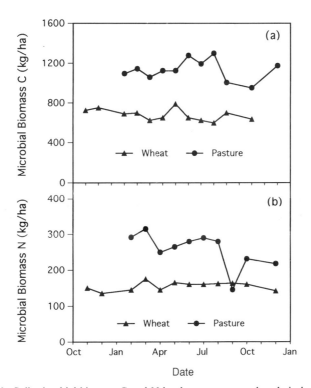

Fig. 8-10. Soil microbial biomass C and N levels over an annual cycle in long-term pasture and wheat fields (from Patra et al., 1990).

the pasture in December is almost 100 kg N ha^{-1} lower than it is in June and July. A change of this magnitude would be agronomically significant. The temporal patterns of soil MBN and MBC do not, however, match well, which could signify a changing C/N ratio of the biomass, or perhaps this level of analysis is not justified given the possible limitations of the fumigation methods of measuring microbial biomass. On the other hand, the soil MBN level in the pasture was generally much higher than that in the wheat field. Perhaps it is this level of analysis that is appropriate for predicting N mineralization. If so, earlier comments about obtaining representative MBN values also apply here.

Insufficient work has so far been done with ASN to provide a good picture of its temporal dynamics. Almost no temporal variability in ASN was observed between May and November 1988 in a silt loam soil cropped to grass or wheat, and ASN levels were considerably higher in the former (Duxbury et al., 1991, and unpublished results). In contrast, declines in ASN of about 15 mg N kg^{-1} soil were observed between May and mid-September 1990 in the plow layer at two of four sites cropped to corn, with no change found at the other two sites. Sharper declines of 40 to 70 mg N kg^{-1} soil were found at the same sites in 1991, which was a drought year (Nkambule, 1993).

It also is important to know whether MBN or ASN pool size predict N availability to crops growing in the field. In making this appraisal, it must be recognized that the methods can at best only predict N supply and not crop recovery of N. When testing for relationships between these measurements and crop N yield, only data where N availability is the major factor controlling crop growth, and where leaching or denitrification losses of mineralized N are unimportant, can be used. Moreover, the dynamics of N release may also influence crop recovery of mineralized N, yet there is reason to doubt that the controls on N mineralization are entirely predictable under field conditions.

A comparative study of a number of potential soil test N methods for corn is underway in New York. In the 1 of 2 yr where the data can be used (drought stress was a major yield determinant in the other year), the best relationship to relative crop yield in unfertilized control plots was obtained with N mineralized in 29 wk ($r^2 = 0.64$; the relationship with calculated PMN values was slightly worse). Therefore, it would seem that the utility of the PMN measurement as an empirical soil test, rather than using it to calculate N mineralization in a theoretical manner, should be investigated further. Neither MBN determined by fumigation–extraction or ASN were good predictors of relative yield. It should be noted, however, that MBN and ASN were not related to PMN in this study, whereas such relationships are often found (see following section).

RELATIONSHIPS BETWEEN POTENTIALLY MINERALIZABLE NITROGEN, MICROBIAL BIOMASS NITROGEN, AND ACTIVE SOIL ORGANIC NITROGEN

The relationship between MBN and PMN was evaluated using eight data sets (see Fig. 8–11 for references). Microbial biomass N was determined by

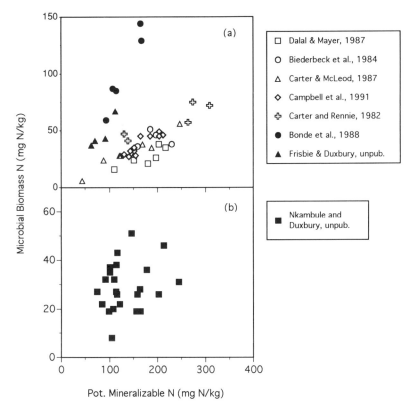

Fig. 8–11. Relationships between potentially mineralizable N and microbial biomass N in various studies using (*a*) fumigation–incubation and (*b*) fumigation–extraction methods of measuring MBN.

the $CHCl_3$ fumigation–incubation procedure in seven of the eight studies, and these data were analyzed together. In the eighth study, MBN was determined by fumigation–extraction, and this data set was analyzed separately. Potentially mineralizable N values were not determined in two of the seven studies where MBN was measured by fumigation–incubation, because the pattern of N mineralization did not lend itself to analysis by the first-order model. These two data sets also were separated out.

When linear regressions were run with the original data from the seven studies measuring MBN by fumigation–incubation, it became apparent that relationships between MBN and PMN varied by study. A possible reason for this was the use of different factors (k_N) to convert the observed N flush to MBN values. When the MBN data were normalized by using the same conversion factor ($k_N = 0.68$), the relationship shown in Fig. 8–11a (open symbols) was found for the five data sets where PMN was calculated. The normalization of data was, however, incomplete because there were other differences in the methods—air-dried soil was used in one of the PMN measurements, MBN was determined after the PMN incubation in one case, and

Table 8-2. Relationships between MBN, PMN, and ASN measurements.

Parameters	No. of data sets/soils	n	r^2	Equation
PMN vs. MBN	5‡/20	34	0.72	PMN = 3.27MBN + 49
N_m 17/23† vs. MBN	2‡/2	9	0.93	N_m 17/23 = 0.92MBN + 39
N_m 29 vs. MBN	1‡/10	23	0.05	
N_m 17 vs. ASN	1§/1	16	0.87	ASN = 1.54 N_m 17 + 0.05
N_m 29 vs. ASN	1¶/10	23	0.25	
ASN vs. MBN	1§/1	16	0.73	ASN = 2.11MBN + 0.08
ASN vs. MBN	1#/10	23	0.80	ASN = 5.56 MBN + 40

† Cumulative N mineralized in the number of weeks indicated.
‡ Data from sources indicated in Fig. 8-11.
§ Unpublished data from S.H. Frisbie and J.M. Duxbury, 1990.
¶ Nkambule, 1993.
# Data from Fig. 8-12.

different controls (0–10 or 10–20 d) were used in MBN determinations. Nevertheless, the relationship is good enough ($r^2 = 0.72$) to make a plea for standardization of methods so that results of different investigations can be directly compared in the future. An excellent relationship ($r^2 = 0.93$) also was found for the other two data sets where N mineralized in 17 (S.H. Frisbie et al., 1990, unpublished data) and 23 (Bonde et al., 1988) weeks of incubation at 30 °C was used instead of PMN values (Fig. 8-11a; closed symbols).

In contrast, no relationship was obtained between PMN (or N mineralized in 29 wk) and MBN in a study involving 10 agricultural sites in New York where MBN was determined by the fumigation–extraction method (Fig. 8-11b). The reason for this result is unclear, but we assume that it is related to the method of determining MBN, because a good relationship with PMN was found in another study in New York ($r^2 = 0.93$) where MBN was determined by the fumigation–incubation method (see Frisbie et al. data in Fig. 8-11a and Table 8-2). In this study, which was carried out at a single site, tillage and soil depth were variables. A possible cause of these differences is that the conversion factor (k_N) for the fumigation–extraction method of measuring MBN varies with soils. Similarly, a good relationship was found between ASN and N mineralized in a long-term incubation (used instead of PMN) in the study at the single site ($r^2 = 0.87$) but not in the study involving 10 sites ($r^2 = 0.25$). The reason for this difference is not known.

Relationships between MBN and ASN ($r^2 = 0.73$–0.80) were found for all of the studies carried out in New York, regardless of the method of measuring MBN (Fig. 8-12). However, the relationships differed quantitatively depending on the method of measuring microbial biomass (Table 8-2). This result again indicates nonconformity between the fumigation–incubation and fumigation–extraction methods of measuring MBN. Overall, the results of the studies in New York are not entirely consistent, and further research is needed to resolve why this is so.

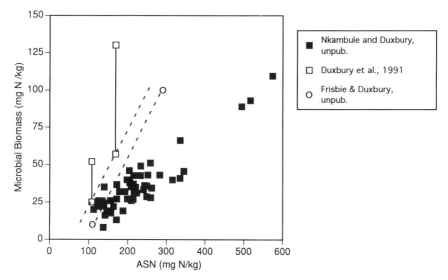

Fig. 8–12. Relationships between active soil N and microbial biomass N in various soils from New York using fumigation–incubation (open symbols) and fumigation–extraction (closed squares) methods of measuring MBN; vertical lines joining the open squares show the range of values obtained with different methods of calculating MBN (the lowest values are comparable to the other data).

CONCLUSIONS AND RECOMMENDATIONS

Although it is clear that more work on methods of measuring soil microbial biomass N and active soil N pools is needed, we are nevertheless making progress toward the goal of measuring the pools of organic N commonly defined in models of organic matter dynamics. Research on methods should be supplemented with rigorous expression of uncertainties associated with the measurements; this is especially needed for microbial biomass measurements.

All three parameters, MBN, PMN, and ASN, show promise as a measure of soil quality, although more research is needed to better define how they should be used and the limits of their use. These measurements also may ultimately prove valuable as predictors of soil N supply, but at this time there are unresolved inconsistencies in data addressing this question. It may, however, be impossible to predict actual N mineralization in the field if unpredictable environmental events, e.g., drying and rewetting of soil, prove to be a major control on mineralization.

Agreement on standardized procedures for measuring MBN and PMN would greatly help efforts to understand relationships between different active soil N measurements and to develop general principles of soil science in this area.

More attention must be given to why measurements of MBN and active soil N pools are being made. As pointed out by Smith and Paul (1990), the titles of something like 90% of all published papers dealing with microbial biomass measurements are either "Effect of . . . on soil microbial biomass" or "The relationship between . . . and soil microbial biomass." We must replace what appears to be measurement for measurements sake with meaningful questions about the functioning of soil–plant systems that these measurements can help answer. An example is the often cited role of the microbial biomass as a reservoir of nutrients for plant growth, yet there is not a single paper that convincingly supports this hypothesis.

ACKNOWLEDGMENTS

We thank Francisco Costa, Seth Frisbie, Diana Friedman, Jean Fruci, and Dean Linscott for permission to use unpublished data, and, together with Julie Lauren, for stimulating discussion on this topic and manuscript.

REFERENCES

Addiscott, T.M. 1983. Kinetics and temperature relationships of mineralization and nitrification in Rothamsted soils with different histories. J. Soil Sci. 34:343-353.

Agarawal, A.S., B.R. Singh, and Y. Kanheiro. 1971. Soil nitrogen and carbon mineralization as affected by drying–rewetting cycles. Soil Sci. Soc. Am. Proc. 35:96-100.

Anderson, J.P.E., and K.H. Domsch. 1978. Mineralization of bacteria and fungi in chloroform fumigated soils. Soil Biol. Biochem. 10:207-213.

Azam, F., R.L. Mulvaney, and F.J. Stevenson. 1988. Determination of in situ k_N by the chloroform incubation method and mineralization of biomass N under anaerobic conditions. Plant Soil 111:87-93.

Beauchamp, E.G., W.D. Reynolds, D. Brasche-Villeneuve, and K. Kirby. 1986. Nitrogen mineralization kinetics with different soil pretreatments and cropping histories. Soil Sci. Soc. Am. J. 50:1478-1483.

Biederbeck, V.O., C.A. Campbell, and R.P. Zentner. 1984. Effect of crop rotation and fertilization on some biological properties of a loam in Southwestsern Saskatchewan. Can. J. Soil Sci. 64:355-367.

Birch, H.F. 1958. The effect of soil drying on humus decomposition and nitrogen availability. Plant Soil 10:9-31.

Bonde, T.A., and T. Lindberg. 1988. Nitrogen mineralization in soil during long-term aerobic laboratory incubations: A case study. J. Environ. Qual. 17:414-417.

Bonde, T.A., and T. Rosswall. 1987. Seasonal variation of potentially mineralizable nitrogen in four cropping systems. Soil Sci. Soc. Am. J. 51:1508-1514.

Bonde, T.A., J. Schnurer, and T. Rosswell. 1988. Microbial biomass as a fraction of potentially mineralizable nitrogen in soils from long term field experiments. Soil Biol. Biochem. 20:447-452.

Brookes, P.C., A. Landeman, G. Pruden, and D.S. Jenkinson. 1985. Chloroform fumigation and the release of soil N: A rapid direct extraction method to measure microbial biomass N in soil. Soil Biol. Biochem. 17:837-842.

Burton, D.L., and W.B. McGill. 1991. Spatial and temporal fluctuation in biomass, nitrogen mineralizing reactions and mineral nitrogen in a soil cropped to barley. Can. J. Soil Sic. 72:31-42.

Cabrera, M.L., and D.E. Kissel. 1988a. Potentially mineralizable nitrogen in disturbed and undisturbed soil samples. Soil Sci. Soc. Am. J. 52:1010-1015.

Cabrera, M.L., and D.E. Kissel. 1988b. Evaluation of a method to predict nitrogen mineralized from soil organic matter under field conditions. Soil Sci. Soc. Am. J. 52:1027-1031.

Campbell, C.A., V.O. Biederbeck, R.P. Zentner, and G.P. Lafond. 1991. Effect of crop rotations and cultural practices on soil organic matter, microbial biomass and respiration in a thin black Chernozem. Can. J. Soil Sci. 71:363–376.

Carter, M.R., and J.A. McLeod. 1987. Biological properties of some Prince Edward Island soils: Relationship between microbial nitrogen and mineralizable nitrogen. Can. J. Soil Sci. 67:333–340.

Carter, M.R., and D.A. Rennie. 1982. Changes in soil quality under zero tillage farming systems: Distribution of microbial biomass and mineralizable C and N potentials. Can. J. Soil Sci. 62:587–597.

Cassman, K.G., and D.N. Munns. 1980. Nitrogen mineralization as affected by soil moisture, temperature, and depth. Soil Sci. Soc. Am. J. 44:1233–1237.

Craswell, E.T., and S.A. Waring. 1972. Effect of grinding on the decomposition of soil organic matter: I. Effect on the mineralization of nitrogen in relation to soil type. Soil Biol. Biochem. 4:427–433.

Dalal, R.C., and R.J. Mayer. 1987. Long term trends in fertility of soils under continuous cultivation and cereal cropping in Southern Queensland: VII. Dynamics of nitrogen mineralization potentials and microbial biomass. Aust. J. Soil Res. 25:461–472.

Deans, J.R., J.A.E. Molina, C.E. Clapp. 1986. Models for predicting potentially mineralizable nitrogen and decomposition constants. Soil Sci. Soc. Am. J. 50:323–326.

Doran, J. 1987. Microbial biomass and mineralizable nitrogen distributions in no-tillage and plowed soils. Biol. Fertil. Soils 5:68–75.

Duxbury, J.M., J.G. Lauren, and J.R. Fruci. 1991. Measurement of the biologically active soil nitrogen fraction by a ^{15}N technique. Agric. Ecosyst. Environ. 34:121–129.

Follett, R.F., and D.S. Schimel. 1989. Effect of tillage practices on microbial biomass dynamics. Soil Sci. Soc. Am. J. 53:1091–1096.

Friedman, D.B. 1993. Carbon, nitrogen, and aggregation dynamics in low-input and reduced tillage cropping systems. M.S. thesis. Cornell Univ., Ithaca, NY.

Herlihy, M. 1979. Nitrogen mineralization in soils of varying texture, moisture, and organic matter. Plant Soil 53:255–267.

Houot, S., J.A.E. Molina, R. Chaussad, and C.E. Clapp. 1989. Simulation by NCSOIL of net mineralization in soils from the Deherain 36 Parcelles plots at Grignon. Soil Sci. Soc. Am. J. 53:451–455.

Jansson, S.V. 1958. Tracer studies on nitrogen transformations in soil with special attention to mineralization–immobilization relationships. Ann. R. Agric. Coll. Swed. 24:101–361.

Jenkinson, D.S. 1976. The effects of biocidal treatments on metabolism in soil: IV. The decomposition of fumigated organisms in soil. Soil Biol. Biochem. 8:203–208.

Jenkinson, D.S. 1988. Determination of microbial biomass carbon and nitrogen in soil. p. 368–386. In J.R. Wilson (ed.) Advances in nitrogen cycling in agricultural ecosystems. CAB International, Wallingford, England.

Juma, N.G., and E.A. Paul. 1984. Mineralizable soil nitrogen: Amounts and extractability ratios. Soil Sci. Soc. Am. J. 48:76–80.

Juma, N.G., E.A. Paul, and B. Mary. 1984. Kinetic analysis of net nitrogen mineralization in soil. Soil Sci. Soc. Am. J. 48:753–757.

Liebhardt, W.C., R.W. Andrews, M.N. Culik, R.R. Harwood, R.R. Janke, J.K. Radhke, and S.L. Rieger-Schwartz. 1989. A comparison of crop production in conventional and low-input cropping systems during the initial conversion to low-input methods. Agron. J. 81:150–159.

Marion, M.G., J. Kummerow, and P.C. Miller. 1981. Predicting nitrogen mineralization in chaparral soils. Soil Sci. Soc. Am. J. 45:956–961.

McGill, W.B., K.R. Canon, J.A. Robertson, and F.D. Cook. 1986. Dynamics of soil microbial biomass and water-soluble organic C in Breton L after 50 years of cropping to two rotations. Can. J. Soil Sic. 66:1–19.

Nkambule, S.V. 1993. A field evaluation of the utility of several methods of estimating plant available soil nitrogen for maize. Ph.D. thesis. Cornell Univ., Ithaca, NY.

Page, A.L., R.H. Miller, and D.R. Keeney (ed.). 1982. Methods of soil analysis. Part 2. 2nd ed. Agron. Monogr. 9. ASA and SSSA, Madison, WI.

Parkinson, D., and E.A. Paul. 1982. Microbial biomass. p. 821–830. In A.L. Page et al. (ed.) Methods of soil analysis. Part 2. 2nd ed. Agron. Monogr. 9. ASA and SSSA, Madison, WI.

Patra, D.D., P.C. Brookes, K. Coleman, and D.S. Jenkinson. 1990. Seasonal changes of soil microbial biomass in an arable and a grassland soil which have been under uniform management for many years. Soil Biol. Biochem. 22:739–742.

Paustian, K., and T.A. Bonde. 1987. Interpreting incubation data on nitrogen mineralization from soil organic matter. INTECOL Bull. 15:101–112.

Rice, C.W., J.H. Grove, and M.S. Smith. 1987. Estimating soil net nitrogen mineralization as affected by tillage and soil drainage due to topographic position. Can. J. Soil Sci. 67:513–520.

Richter, J., A. Nuske, U. Bohmer, and J. Wehrmann. 1980. Simulation of nitrogen mineralization and transport in Loess-Parabrownearthes; plot experiments. Plant Soil 54:329–337.

Schmidt, E.L., and E.A. Paul. 1982. Microscopic methods for soil microorganisms. p. 803–814. In A.L. Page et al. (ed.) Methods of soil analysis. Part 2. Agron. Monogr. 9. ASA and SSSA, Madison, WI.

Shen, S.M., G. Pruden, and D.S. Jenkinson. 1984. Mineralization and immobilization of nitrogen in fumigated soil and the measurement of microbial biomass nitrogen. Soil Biol. Biochem. 16:437–444.

Sierra, J. 1990. Analysis of soil nitrogen mineralization as estimated by exponential models. Soil Biol. Biochem. 22:1151–1153.

Smith, J.L., and E.A. Paul. 1990. The significance of soil microbial biomass estimations. p. 357–396. In J.M. Bollag and G. Stotzky (ed.) Soil biochemistry. Vol. 6. Marcel Dekker, New York.

Smith, S.J., L.B. Young, and G.E. Miller. 1977. Evaluation of soil nitrogen mineralization potentials under modified field conditions. Soil Sci. Soc. Am. 41:74–76.

Stanford, G., J.N. Carter, D.H. Schwaninger, and J.J. Meisinger. 1977. Residual nitrate and mineralizable soil nitrogen in relation to nitrogen uptake by irrigated sugar beets. Agron. J. 69:303–308.

Stanford, G., J.N. Carter, and S.J. Smith. 1974. Estimates of potentially mineralizable soil nitrogen based on short-term incubations. Soil Sci. Soc. Am. Proc. 38:99–102.

Stanford, G., and E. Epstein. 1974. Nitrogen mineralization–water relations in soils. Soil Sci. Sco. Am. Proc. 38:103–107.

Stanford, G., M.H. Frere, and D.H. Schwaninger. 1973. Temperature coefficients of soil nitrogen mineralization. Soil Sci. 115:321–323.

Stanford, G., and S.J. Smith. 1972. Nitrogen mineralization potentials of soils. Soil Sci. Soc. Am. Proc. 36:465–472.

Stanford, G., and S.J. Smith. 1978. Oxidative release of potentially mineralizable soil nitrogen by acid permanganate extraction. Soil Sic. 126:210–218.

Tabatabai, M.A., and A.A. Al-Khafaji. 1980. Comparison of nitrogen and sulfur mineralization in soils. Soil Sci. Soc. Am. J. 44:1000–1006.

Talpaz, H., P. Fine, and B. Bar-Yosef. 1981. On the estimation of N-mineralization parameters from incubation experiments. Soil Sci. Soc. Am. J. 45:993–996.

Tisdall, J.M., and J.M. Oades. 1982. Organic matter and water stable aggregates in soils. J. Soil Sci. 33:141–163.

Verhoef, H.A., and L. Brussaard. 1990. Decomposition and nitrogen mineralization in natural and agroecosystems: The contribution of soil animals. Biogeochem. 11:175–211.

Verstraete, W., and J.P. Voets. 1976. Nitrogen mineralization tests and potentials in relation to soil management. Pedologie 26:15–26.

Voroney, R.P., and E.A. Paul. 1984. Determination of k_C and k_N in situ for calibration of the chloroform fumigation-incubation method. Soil Biol. Biochem. 16:9–14.

Wardle, D.A., and D. Parkinson. 1990. Interactions between microclimatic variables and the soil microbial biomass. Biol. Fertil. Soils 9:273–280.

Chapters
9–18

Extended Summaries

Chapters 9–18

9 Multiple Variable Indicator Kriging: A Procedure for Integrating Soil Quality Indicators

Jeffrey L. Smith, Jonathan J. Halvorson, and Robert I. Papendick

USDA-ARS
Washington State University
Pullman, Washington

In the early 1990s there has been an increasing awareness of the importance of soil quality for agricultural sustainability. This awareness has gained attention worldwide and has prompted several symposia on soil quality. Much of the effort in these conferences has been directed toward developing a definition of soil quality. The phrases and terms surrounding the concept of a high-quality soil relate to its ability to supply crop nutrients (productivity), resist erosion, provide habitat to microorganisms and microfauna, and ameliorate environmental pollution.

There are several working groups within the USA attempting to identify the parameters that would serve as long-term indicators of soil quality. Within each group, researchers, representing several of the soil subdisciplines (i.e., microbiology, chemistry, etc.) have been asked to furnish criteria that should be included in the overall assessment of soil quality. The list of parameters proposed to assess soil quality is broad and includes physical, chemical, and biological parameters as well as descriptive variables. For a given location, measurements of these soil quality indicators will be combined into a multiple variable data set to define local soil quality or an index thereof.

We have developed an approach to integrate an unlimited number of soil quality indicators into an overall soil quality indicator or index. We have termed this procedure *multiple variable indicator transform* (MVIT), which transforms measured data values into a soil quality index according to specified criteria. The criteria, developed independently for each indicator, are critical values or ranges of values that represent the best estimate of good soil quality. These soil quality criteria and critical values can be developed and compared on a regional basis. The MVIT data can be used as an index of good and poor quality soil where soil areas are rated on a defined scale;

however, it may be more useful to evaluate soil quality on a landscape basis as the probability of areas having good soil quality.

To determine the probability of an area having good soil quality, we have utilized an approach based on nonparametric geostatistics called *indicator kriging* (IK), which uses the transformed data from MVIT to estimate values for locations that have not been sampled. Through the estimation procedure, maps are developed that indicate the probability of meeting a soil quality criteria on the landscape level. Because we are integrating several variables, we term this procedure *multiple variable indicator kriging* (MVIK). The method is capable of defining soil parameter gradients, boundaries, and interactive effects through integration of criteria developed by soil quality experts. Thus, this method does not define soil quality. It only integrates the criteria chosen to represent good soil quality.

SOIL QUALITY DATA SET

The soil samples for this study were taken from a slightly north-sloping fallow wheat (*Triticum aestivum* L.) field northwest of Pullman, WA. The soil is a palouse silt loam classified as a Pachic Ultic Haploxeroll. A total of 220 soil samples were taken over approximately 0.5 ha (50 by 110 m). The sampling design was on a regular 10 × 10 m grid with smaller sampling distances (lags) randomly placed throughout the larger grid. Soil samples were taken from the surface 0 to 5 cm and stored at 4 °C until analysis. Soil samples were analyzed for pH, total inorganic nitrogen (TIN) (where $NH_4 + NO_3$ = TIN), and electrical conductivity using standard methods of analysis (Page, 1982). Soil pH and conductivity were measured on a 1:1 soil weight/water weight mixture.

Table 9-1 shows the summary statistics for the three soil parameters. Both the conductivity and TIN parameters are positively skewed and tended to be log normal, even though the medians were close to the mean values. The pH distribution was normal, with a low coefficient of variation and lit-

Table 9-1. Summary statistics for three soil indicator parameters.

Statistic	Soil indicator parameter		
	Conductivity	pH	TIN
	dS m^{-1}		mg kg^{-1}
Number of points	220	220	220
Mean	2.13	5.14	22.4
Standard deviation	0.68	0.24	15.7
Coefficient of variation (%)	32.0	4.69	70.3
Skewness	1.79	0.12	1.88
Minimum	0.95	4.56	2.64
25th percentile	1.68	4.96	10.5
Median	2.00	5.12	18.3
75th percentile	2.45	5.30	29.6
Maximum	5.79	5.74	110.7

tle skewness. This data set was not edited for outliers and will be used as is in the following spatial analysis.

The three soil parameters that we have measured may not be the most important as related to soil quality for this area or land use. They are, however, simple measurements that are routinely performed in soils laboratories. We therefore use this data only to show how multiple soil parameters can be integrated to develop probability indices of acceptable or unacceptable soil quality.

INDICATOR TRANSFORMATION

Indicator transformation of continuous variables into indicator variables is particularly useful for skewed data sets where the estimation of cumulative distributions is desirable (Isaaks & Srivastava, 1989). In addition, the use of indicator transformation makes ordinary kriging a more powerful technique than polygon or inverse distance squared estimation methods. The first step in the procedure is to transform the raw data into indicator data based on whether each datum meets a chosen critical threshold value. The second step is the integration of indicator data using MVIT, which will be discussed below.

To transform the raw data of pH, TIN and conductivity, critical threshold values must be chosen. The critical threshold values can be arbitrary, based on scientific evidence or derived from experience. These values should relate to the pertinent soil quality question. If, for example, soil quality is based on crop production, the critical values may be chosen such that pH must be greater than 4.0, conductivity between 2.0 and 5.0 dS m^{-1} and TIN greater than 5.0 mg kg^{-1}. Keep in mind that all of these critical thresholds could represent ranges of values or lower acceptable limits.

We code the data as *0* if it does not meet the chosen criteria or a *1* if it meets the predetermined threshold. In our example data set, for simplicity, we have chosen the median as the critical value that each raw data value has to be above. For this analysis, soil conductivity was considered a measure of soiluble ionic nutrients since all of the values were below 7 dS m^{-1} where salinity may affect wheat growth. Thus, conductivities above the median were suggestive of adequate soluble plant nutrients. Table 9–2 gives an example indicator transformation of data from three locations. For the "5,5" meter location, both conductivity and pH are below the median value (Table 9–1) and thus are coded 0, whereas TIN is greater than the median and is coded 1. All data from each of the 220 locations is transformed in the same manner.

MULTIPLE VARIABLE INDICATOR TRANSFORM

The second operation necessary is the integration of all the measured parameters that were indicator transformed in the previous step. To do this,

Table 9-2. Indicator transform of data from three example locations using the median as the critical value.

Location	Conductivity	pH	TIN	Indicator value		
				Conductivity	pH	TIN
m	dS m^{-1}		mg kg^{-1}			
5, 5	1.73	4.93	20.7	0	0	1
20, 10	2.16	5.32	16.0	1	1	0
95, 5	2.69	5.37	27.3	1	1	1

the number of soil quality parameters that must meet the threshold criteria to be deemed *good* soil quality must be defined. If soil quality criteria is based on three parameters, do all three parameters need to meet their critical threshold for the soil at that location to be coded *good* soil. *Good soil quality* might be defined as those locations that meet all (three of three) or only some (one of three, or two of three) of the threshold criteria. Our example is based on three parameters. Soil quality, however, could be based on any number of parameters; thus, those needing to meet the chosen criteria could be any number of parameters or combination of parameters (Smith et al., 1993).

For this example we will develop three integration scenarios termed COMB1, COMB2, and COMB3, which require that 1 of 3, 2 of 3, and 3 of 3 soil parameters, respectively, must meet its criteria value to be coded good soil quality. Table 9-3 shows the integration of individual parameter indicator values that meet or fail to meet specified threshold criteria. The integrated COMB1 scenario shows that all three locations met the requirement that at least one soil quality parameter met its criteria. For the COMB2 scenario, two locations met the requirement that at least two parameters met specified criteria. The COMB3 shows only one location with acceptable soil quality based on requiring all three parameters to meet specific threshold values. This is somewhat intuitive since we are being more restrictive in the requirements from COMB1 to COMB3, respectively. Once the indicator values for each location are integrated into a combined indicator value, we can use MVIK to produce landscape level maps.

Table 9-3. Integration of indicator transformed data from soil parameters into a single combined indicator value for three different integration scenarios.

Location	Indicator value			Combined indicator value		
	Conductivity	pH	TIN	COMB1	COMB2	COMB3
m						
5, 5	0	0	1	1	0	0
20, 10	1	1	0	1	1	0
95, 5	1	1	1	1	1	1

VARIOGRAPHY

The semivariance and its plot with distance, the variogram, are the traditional tools used in geostatistics. However, the variogram can be an incomplete source of information about the spatial continuity of a soil property if the data are skewed, clustered, or if unequal lag means or lag variances occur (i.e., when the assumption of stationarity is inappropriate). The nonergotic autocorrelation function accounts for changes in both lag means and variances.

$$p^*(h) = \frac{\frac{1}{N(h)} \sum_{i=1}^{N(h)} [z(x_i) - m_{-h}][z(x_i + h) - m_{+h}]}{s_{-h} \, s_{+h}} \tag{1}$$

where $z(x_i)$ and $Z(x_i + h)$ are two data points separated by the vector h. Datum $z(x_i)$ is the tail and $z(x_i + h)$ is the head of the vector, $N(h)$ are the total number of data pairs separated by vector h, m_{-h} and m_{+h} are the means of the points that correspond to the tail and the head of the vector, respectively, and s_{-h} and s_{+h} are the standard deviations of the tail and head values of the vector, respectively (Rossi et al., 1992).

Figure 9-1 shows an idealized plot of the autocorrelogram function with lag distance (correlogram). Three properties of the graph are the nugget, sill, and range. The *nugget* is the amount of variance not explained or modeled as spatial correlation. It is the apparent ordinate intercept and is due to unsampled correlation below the smallest lag and to experimental error. The *sill*, if present, is characterized by a leveling-off of the correlogram model. It indicates that spatial correlation, on the average, is constant. The lag value when the correlogram model reaches the sill is the *range*. It represents the maximum separation distance within which samples are spatially correlated.

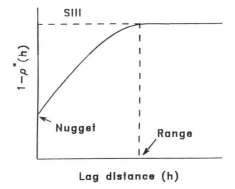

Fig. 9-1. Idealized plot of the autocorrelation function (Eq. [1]) with lag distance, usually termed *correlogram*, showing the nugget, sill, and range of the function. Values calculated from Eq. [1] are subtracted from one, so the graph is in standard variogram format.

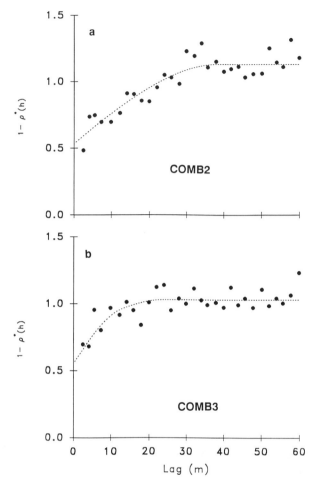

Fig. 9-2. Indicator correlograms of the integrated parameters for COMB2 and COMB3, and the fitted positive definite models (*dashed lines*). Values calculated from Eq. [1] are subtracted from one, so the graph is in standard variogram format.

Figure 9-2 shows the plot of the lag correlation-coefficient values vs. distance, or the indicator correlogram, for COMB2 and COMB3. Both variables show spatial structure with the range for COMB2 being 35 m and for COMB3 being 15 m. The shorter range observed for the correlogram of COMB3 may be related to differences in the joint probabilities of meeting three of three vs. two of three of the threshold criteria at each location. The nugget or autocorrelation at lag 0 defines the proportion of parameter variation that is unexplained by spatial dependence. The relatively large nugget for both indicator correlograms (Fig. 9-2) may be due to low or negative correlations between measured parameters and the subsequent requirement that these parameters both be above specific critical values.

MULTIPLE VARIABLE INDICATOR KRIGING

The models in Fig. 9-2 were used in an ordinary kriging procedure to estimate indicator values at unsampled locations. These numbers correspond to the probability that the unknown values meet or do not meet the specified threshold (Journel, 1988). For the combined transformed data we used ordinary block kriging to estimate unknown values of 2.5 by 2.5 m cells. Each discretized point was estimated using the information from all neighbors within a 20-m search radius and from the combined indicator correlogram.

Figure 9-3 shows the probability plots for COMB2 (Fig. 9-3a) and COMB3 (Fig. 9-3b). The contours indicate the probability that the area enclosed will meet the chosen criteria. For the COMB2 scenario (Fig. 9-3a) the contour value at any location is the probability that two of three soil properties meet their critical threshold values.

Casual observation of Fig. 9-3a shows an area that has a 90% probability of having at least two parameters meeting a criteria, and large areas with 50% probabilities of meeting soil quality requirements. However, in Fig. 9-3b the 90% area disappears, and most of the area shows 10% or less compliance. Since we have tightened our requirements from COMB2 to COMB3 this result is not surprising. It does, however, point out that the

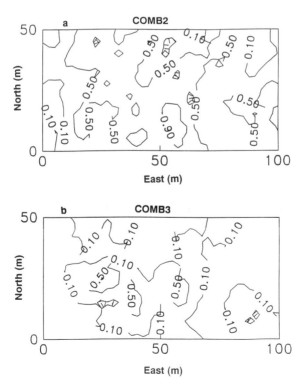

Fig. 9-3. Indicator kriged contour probability plots for (*a*) COMB2 and for (*b*) COMB3.

criteria development phase for soil quality evaluation procedures will be very important.

The contour maps presented could be easily converted via the geographic information system (GIS) to land areas having acceptable soil quality and those areas needing management. Multiple variable indicator kriging can be used to monitor soil quality or improvement over time, and across large areas if enough samples can be taken. In this example enough samples were taken to adequately develop and test the procedure and may exceed the actual number required.

SUMMARY AND CONCLUSIONS

Soil quality is the most important factor for sustaining the worldwide biosphere. To quantify soil quality, specific soil indicators need to be measured spatially. These indicators are mainly soil properties whose values relate directly to soil quality but may also include policy, economic, or environmental considerations. Because asessing soil quality is complex, several individual soil quality indicators need to be integrated to form a soil quality index. This integration needs to be flexible enough to evaluate soil quality at spatial scales ranging from the farm to the regional level. We have developed a MVIK procedure that provides a means of integrating soil quality parameters into an index to produce soil quality maps on a landscape basis. These maps indicate the areas on a landscape that have a high probability of having *good* soil quality according to predetermined criteria.

It is our opinion that the use of the MVIK procedure to produce probability maps is superior to rating soil quality on an index scale of 1 to 10. This is because there is more flexibility to incorporate management decisions and economic and environmental constraints into the analysis. In addition, by evaluating the individual indicator values, it is possible to identify specific soil properties that affect soil quality most strongly. The following steps outline the MVIT and MVIK procedure.

1. Identify and measure soil quality parameters.
2. Transform individual parameter values to indicator values (MVIT) based on critical thresholds.
3. Integrate multiple indicator values (MVIT) based on the number of parameters that must meet the chosen criteria for that parameter.
4. Use the MVIT values in a kriging procedure (MVIK).
5. Produce kriged contour maps of the probability of soil meeting the chosen criteria.

REFERENCES

Journel, A.G. 1988. Non-parametric geostatistics for risk and additional sampling assessment. p. 45–72. *in* L. Keith (ed.) Principles of environmental sampling. American Chemical Society, Washington, DC.

Isaaks, E.H., and R.M. Srivastava. 1989. An introduction to applied geostatistics. Oxford Univ. Press, New York.

Page, A.L. (ed.). 1982. Methods of soil analysis. Part 2. 2nd ed. Agron. Monogr. 9. ASA and SSSA, Madison, WI.

Rossi, R.E., D.J. Mulia, A.G. Journel, and E.H. Franz. 1992. Geostatistical tools for modeling and interpreting ecological spatial independence. Ecol. Monogr. 62:277–314.

Smith, J.L., J.J. Halvorson, and R.I. Papendick. 1993. Using multivariable indicator kriging for evaluating soil quality. Soil. Sci. Soc. Am. J. 57:743–749.

10 Descriptive and Analytical Characterization of Soil Quality/Health[1]

M. J. Garlynd, D. E. Romig, and R. F. Harris

University of Wisconsin
Madison, Wisconsin

A. V. Kurakov

Moscow State University
Moscow, Russia

National and worldwide initiatives show development and application of methodology for assessment and monitoring of *soil quality* (term favored by scientists)/*soil health* (term favored by farmers) as an emerging high priority to provide a basis for maintenance and remediation of soil resources (Larson & Pierce, 1991; Arshad & Coen, 1992; Haberern, 1992; Papendick & Parr, 1992). *Soil health* is more a farmer than an institution concept, stemming from farmer concerns and observational experiences in the field. Management strategies of soil quality/health and meaningful communication between diverse farmer, agribusiness, and academic interest groups require the availability of mutually respected soil quality/health assessment tools.

The Wisconsin Soil Quality/Health program at the University of Wisconsin-Madison was initiated in 1990 in response to farmer concern expressed at statewide "Listening Meetings" on the need for university work on soil biological quality and soil health. A key working hypothesis is that farmer observational knowledge on soil health can and should be integrated with scientist analytical expertise in soil quality. Farmer–scientist partnership (Porter, 1991; Harris, 1992; Harris et al., 1992; Porter et al., 1992; and Chapt. 2, this book) has resulted in a coordinated set of conceptual, information gathering and analysis, and management tools for soil quality/health assess-

[1] Supported by the University of Wisconsin (UW) Center for Integrated Agricultural Systems and the Nutrient and Pest Management Program, the UW Agricultural Technology and Family Farm Institute, the Wisconsin Integrated Cropping Systems Trial and the Kellogg Foundation, Wisconsin Department of Agriculture, Trade, and Consumer Protection (DATCP) Sustainable Agriculture Demonstration Project, the Wisconsin DATCP Fertilizer Research Council, and the Wisconsin Liming Materials Council.

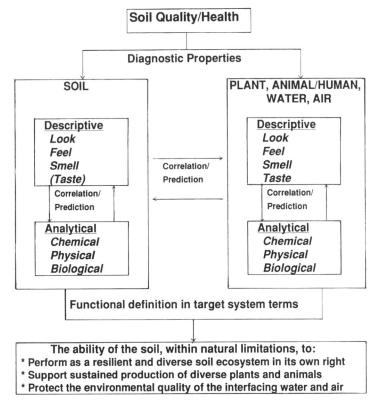

Fig. 10-1. Soil quality/health interpretive framework.

ment. The management tools currently envisaged are: (i) a predictive report card to record and translate soil descriptive and analytical properties into predictive performance of soil, crop, animal–human, water, and air target systems; and (ii) a nontechnical manual based largely on descriptive properties of soil and nonsoil target systems that farmers can use to assess and monitor soil quality/health in their management practice. This chapter focuses on progress on our conceptual tool, the interpretive framework, and our still-evolving interview guide, questionnaires, and correlative report card information-gathering and analysis tools.

Farmer–institution partnership has been essential for the progress made, and ultimate development and validation of soil quality/health management tools necessitate the joint labors of farmers and scientists of diverse expertise, farming interest, and philosophy.

INTERPRETIVE FRAMEWORK

The interpretive framework (Fig. 10–1) recognizes that the target systems for soil quality/health characterization are the soil itself, and the plant, animal–human, water, and air systems supported by and interfacing with soil. For each target system, the descriptive components are grouped into the major sensory categories of look, feel, smell, and taste, and analytical components into classical chemical, physical, and biological properties. Correlative and predictive relationships between diagnostic properties within and between soil and nonsoil target systems are also recognized. It follows that the definition of soil quality/health becomes the capability of soil to (i) function as a resilient and diverse ecosystem in its own right, (ii) to support sustained production of diverse plants and animals, and (iii) to protect the environmental quality of the interfacing water and air (Fig. 10–1).

INTERVIEW GUIDE

The interview guide is an information-gathering and communication tool that is structurally a table form of the interpretive framework, expanded to include provisions for specific properties and qualifying descriptors. In accordance with the interpretive framework, room is also provided for input on correlations within and between the target system diagnostic properties.

The interview guide is currently administered in three phases. The first phase involves informing the respondent that we are interested in how he or she personally recognizes soil quality/health. The question is posed, "How do you personally recognize a high-quality/healthy soil?" Answers are recorded by the interviewer onto the appropriate location in the interview guide, but in such a way that the respondent is not aware of the guide's structure. The interviewer's task is to ensure that properties are clearly defined to the level of word descriptors for descriptive properties and units for analytical properties. During a recap of the responses, the respondent is asked to identify the most important properties characterizing their perception of a high-quality/healthy soil. A similar approach is taken for the question, "How do you personally recognize a low-quality/unhealthy soil?" The data obtained in this phase identify the unbiased viewpoint and the primary concerns of the respondent about soil quality/health.

The second phase is initiated by using the interpretive framework as a communication tool for examination of the partially completed interview guide with the respondent. Each information category not specifically addressed by the respondent in Phase 1 is considered. Any new prompted answers are recorded (in different color) on the guide or are labeled *not important* if they do not contribute to the respondent's recognition of soil quality/health. Finally, correlations of diagnostic properties between and within target systems are solicited using the interpretive framework as a reference base.

The third phase involves documentation of background information on farm location and type, management practice, and farming philosphy.

Pretesting and refinement of the interview guide are currently underway. Data generated from administration of the interview to representatives of specific farming type and philosophy will provide a rich source of indigenous knowledge for interpretation by social scientists and further the development of other information-gathering and management tools.

GENERAL QUESTIONNAIRE

The general questionnaire is under development for use as a self-administered tool for establishment of the language used by farmers and scientists to describe and assess soil quality/health. It will be administered to groups selected on the basis, for example, of geographic region, management practice, and farm type. The general questionnaire is currently composed of four major sections: (i) the diagnostic descriptive and, to a lesser extent, analytical properties characterizing high-quality/health and low-quality/unhealthy soil, proceeding from the plant through the animal–human, water, and air nonsoil target systems to the soil target system; (ii) the relative ranking of the key diagnostic properties; (iii) correlations between and among key diagnostic descriptive and analytical properties of soil and nonsoil target systems; and (iv) respondent background and farming philosophy.

Experience has shown that the most effective way to solicit perceptions is to start with how respondents recognize soil quality/health as a function of diagnostic properties of the nonsoil target systems, and then proceed to diagnostic properties of the soil target system, rather than the other way around. Based on individual farmer and group responses to earlier versions of our interpretive framework (Chapt. 2, this book), the major descriptive categories identified in Fig. 10–1 have been expanded to the level of specific properties and qualifying descriptors. Figure 10–2 is the first page of the current version of our general questionnaire and illustrates the basic approach taken for the diagnostic property section (Section 1) of the questionnaire.

SITE-SPECIFIC QUESTIONNAIRE

The structure of the site-specific questionnaire is similar to that of the general questionnaire, except it reflects the fact there is only one soil involved, and its quality/health status as identified by the farmer may range from high quality/very healthy to low quality/very unhealthy. The site-specific questionnaire is self-administered. Its major function is to obtain in standardized terms the year-integrated rationale of a cooperating farmer for assignment of a specific soil quality/health status to an experimental site chosen for detailed evaluation.

The site-specific questionnaire is composed of the same four major sections as the general questionnaire. Figure 10–3 illustrates the structure of the diagnostic property section (Section 1) and specifically addresses the relative importance and word options for the aggregation property of the descriptive look–feel category of the soil target system.

GENERAL QUESTIONNAIRE FOR SOIL QUALITY/HEALTH

Name: Date:

I. Do you associate the quality/health of a soil with properties of the plants, animals and humans supported by the soil, or the water and air interfacing with the soil?
Please check ☒ your response.
    O YES - Go to question # II below. (Descriptive Properties of Plants)
    O NO   - Go to page 25, question # XIX (Properties of Soil).

(PLANT)

II. Do you associate the quality/health of a soil with descriptive properties of the plants?
Please check ☒ your response.
    O YES - Go to descriptive properties of plants below (Appearance of Plant Roots).
    O NO   - Go to page 9, question # III (Analytical Properties of Plants).

DESCRIPTIVE PROPERTIES

Look

Appearance of Plant Roots

How important is the Appearance of Plant Roots in your recognition of soil quality/health? Check the level of importance in the box to the right. (If zero, go to Appearance of Plant Stems.)

High	Moderate	Low	Zero
O	O	O	O

Describe the appearance of plant roots you associate with soil depending on it's quality/health. Check the circles that correspond to the appearance of plant roots and a soil's particular quality/health status. Add new descriptors if needed (Other).

DESCRIPTOR	SOIL QUALITY/HEALTH HIGH	LOW
Scrawny	O	O
Color spots of roots	O	O
Straight	O	O
Branching	O	O
Limited	O	O
Developed	O	O
Deep rooted row crops	O	O
Low corn brace roots	O	O
Lots of fungal threads	O	O
Few fungal threads	O	O
Lots of fine roots	O	O
Few fine roots	O	O
Lot of nodules on legumes	O	O
Few nodules on legumes	O	O
Other _____	O	O
_____	O	O
_____	O	O

Go to next plant property (Appearance of Plant Stems).

Fig. 10–2. First page of the generalized questionnaire.

CORRELATIVE REPORT CARD

The correlative report card is a set of forms for recording and comparatively evaluating descriptive and analytical data of potential value for characterizing the soil quality/health of a specific experimental site. As for the other information-gathering tools, the basic content of the correlative report card is consistent with the interpretive framework (Fig. 10–1). The site-specific

Aggregation

How important is Aggregation in the recognition of your soil's quality/health status? Check the level of importance in the box to the right.

High	Moderate	Low	Zero
○	☒	○	○

Describe the aggregation of your soil. Check the circles that correspond to your description of the soil aggregation . Add new descriptors if needed (Other).

Go to next soil property
(Quality of Tillage).

DESCRIPTOR

Cloddy	○
Lumpy	○
Crumbly	☒
Chunky	○
Large aggregates	○
Fine aggregates	○
Dusty	○
Different sizes	☒
Same sizes	○
Other _____	○
_____	○
_____	○

Fig. 10-3. Example of question in descriptive property section of the site-specific questionnaire (farmer-identified healthy soil).

questionnaire, or a site-specific form of the interview guide, is used to obtain farmer perceptions of descriptive diagnostic properties. Standard laboratory methods are used for analytical properties, with current emphasis on the soil quality methodology recommendations of the North Central Region Committee on Soil Organic Matter (NCR-59).

The current version of our correlative report card is composed of four sections:

1. Form for year-integrated, key descriptive properties with qualifying word descriptors, recognized by the cooperating farmer as the basis of the perceived soil quality/health status of the site
2. Table for seasonal analytical properties considered by the scientist as a minimum data set for characterizing the soil quality/health status of the site (e.g., Table 10-1)
3. Open-ended form for farmer-assisted, scientist evaluation of correlations between and among descriptive and analytical properties of soil, plant, animal–human, water and air target systems
4. Form for the farmer's soil quality/health ranking rationale, correlative comments, farming philosophy, and farm background (e.g., Fig. 10-4)

Table 10-1 provides an example of the soil analytical component of the correlative report card. As identified in Table 10-1, our current strategy is to sample in the spring, summer, and fall for a number of years to establish reliable trends for analytical properties.

We have applied progressively evolving variations of the correlative soil quality/health report card to nine farm sites for the summer of 1990 through

Table 10-1. Soil analytical property section of the correlative report card (farmer-identified healthy soils).

Farmer's name:		Health/quality ranking: Healthy		Year: 91–92
Soil analytical property	Descriptors (values with units in parenthesis)			
	Fall 1991	Spring 1992	Summer 1992	
Chemical				
pH	7.8	7.7	7.9	($-\log[\text{H}^a]$)
Nitrate-N	10.1	30.2	10.4	(mg kg^{-1})
Ammonium-N	5.1	5.9	3.5	(mg kg^{-1})
Extractable P	3	16	5	(mg kg^{-1})
Exchangeable C	5183	4333	4667	(mg kg^{-1})
Exchangeable M	800	730	727	(mg kg^{-1})
Exchangeable K	203	197	165	(mg kg^{-1})
Cation exchange capacity	33	28	30	(meq 100 g^{-1})
Physical				
Bulk density	0.95	--	--	(g cm^{-3})
Total water holding capacity	59	--	--	(%)
Macroporosity	26	--	--	(%)
Aggregate stability				
(<2.0mm)	18.9	22.1	24.3	(%)
(>0.25mm)	56.7	61.5	66.4	(%)
Electrical conductivity	17	19	16	(mhos × 10^{-5} cm^{-1})
Plant-available water	7	--	--	(%)
Field water holding capacity	34	--	--	(%)
Particle-size distribution				
(sand)	41	--	--	(%)
(silt)	45	--	--	(%)
(clay)	14	--	--	(%)
Total porosity	64	--	--	(%)
Saturated hydraulic conductivity	75.6	--	--	(1 × 10^{-5} m s^{-1})
Biological				
Organic Matter	6.3	5.8	6.2	(%)
Labile C	324	234	1233	(mg C kg^{-1} OD soil)
Microbial biomass (C)	124.8	71.0	62.3	(mg C 100 kg^{-1})
Arginine ammonification	22.9	32.3	19.7	(mg N 10 kg^{-1} h^{-1})
Dehydrogenase	92.2	168.5	171.1	(mg TPF 10 kg^{-1} h^{-1})

Please provide the following information about your farm and field management:

SOIL QUALITY/HEALTH RANKING BY FARMER

1.What is the status of quality/health for this field? | Healthy _ _ _ _ _ _ _ _ _ _ _ _

2. State the main reasons for your soil quality/ health ranking of this field? | Good, intensive biodecomposition to release nutrients, improved appearance and development of crop, increased yield _ _ _ _ _ _

FARMER CORRELATIVE COMMENTS

1. Are there any associations between descriptive and analytical properties inside the soil or non soil systems that you notice about this field? | In plant system, slow development, thin stems, and narrow leaves result in decreased yields of corn _ _ _ _ _ _ _ _ _ _ _ _

2. Are there any associations between non soil system properties and soil properties that you notice about this field? | High level of residue decomposition and biolife activity in soil leads to sufficient nutrient pool and good growth and yield of plant _ _ _ _

3. Are there any associations you can think of connected with this field? | High rate of herbicide application connected with pollution of water surrounding and underlying field _ _ _ _ _ _ _ _ _ _ _ _

FARM BACKGROUND INFORMATION

GENERAL
FARMER NAME: _ _ _ _ _ _ _ _ _ _ _ _ _ _ _
FARM LOCATION: | Watertown, Wisconsin _ _ _ _ _ _ _ _ _
FARM TYPE: | Cash Grain _ _ _ _ _ _ _ _ _ _ _ _
SOIL CLASSIFICATION: | Mortinon Clay _ _ _ _ _ _ _ _ _ _ _
FARMING PHILOSOPHY: | Low Synthetic Input Conventional _ _ _ _ _ _ _ _

SITE-SPECIFIC
ROTATION: | Corn-Bean _ _ _ _ _ _ _ _ _ _ _ _ _
IRRIGATION/DRAINAGE: | Tiled _ _ _ _ _ _ _ _ _ _ _ _ _ _
TILLAGE: | Ridge-Till, Two Rotary Hoeings, Two Cultivations _ _ _ _ _
NUTRIENT PROGRAM: | Starter Fertilizer: 160# NPK 9-23-30; Sidedress: 110# Anhydrous Ammonia _ _ _ _ _ _ _

WEED, PEST AND
DISEASE CONTROL: | Ducel at Half Rate/ Half Pint per Acre; Bandvil Broadcast Application on 8" Band at Full Rate/Two and Half Pints per Acre_ _ _ _ _ _

PERIOD OF PRESENT
MANAGEMENT: | 7 years _ _ _ _ _ _ _ _ _ _ _ _ _ _ _ _
HISTORY OF FIELD: | Continuous Corn _ _ _ _ _ _ _ _ _ _ _ _ _

Fig. 10–4. Farmer soil quality/health ranking, correlative comments, and background information section of the correlative report card.

the fall of 1992 time period. The sites represent empirically selected pairs of conventional and alternative management and/or sites of farmer-identified high-quality/healthy and low-quality/unhealthy soil. Ongoing experiments using the soil quality/health assessment tools close to those described in this chapter have been in progress since fall 1991. Results are under evaluation and will be integrated with fall 1992 data. Among apparent trends are that farmer-identified high-quality/healthy soils, as compared with low-quality/unhealthy soils, tend to show a higher level of residue decomposition and earthworm activity, higher values for saturated hydraulic conductivity and macroporosity, higher levels of dehydrogenase and arginine hydrolysis enzymatic activity, and higher enzymatic activity/soil microbial biomass ratios.

SUMMARY

The status of a coordinated set of conceptual, information-gathering, and management tools for assessment of soil quality/health based on the farmer–scientist partnership reveals that farmers contribute experience-based descriptive knowledge and scientists provide analytical expertise. The conceptual tool is an interpretive framework identifying target systems, descriptive and analytical components and major categories of diagnostic properties, correlative and predictive interrelations of diagnostic properties between and within the target systems, and a functional definition of soil quality/health in target system terms. The target systems are soil, plant, animal–human, water, and air. The major categories are sensory properties of look, feel, smell, and taste for the descriptive component; and chemical, physical, and biological properties for the analytical component. The information-gathering tools are based on the interpretive framework, and include an interview guide, a general and site-specific questionnaire, and a correlative report card. Projected management tools are a predictive report card and a nontechnical manual.

ACKNOWLEDGMENTS

Conceptual contributions from Pamela Porter and other members of the Wisconsin Soil Quality/Health Program Steering Committee are gratefully acknowledged.

REFERENCES

Arshad, M.A., and G.M. Coen. 1992. Characterization of soil quality: Physical and chemical criteria. Am. J. Altern. Agric. 7:25–32.

Haberern, J. 1992. Coming full circle—the new emphasis on soil quality. Am. J. Altern. Agric. 7:3–4.

Harris, R.F. 1992. Developing a soil health report card. Proc. Wisconsin Fert. Aglime Pest Manage. Conf. 31:245–248.

Harris, R.F., M.J. Garwick, P.A. Porter, and A.V. Kurakov. 1992. Farmer/institution partnership in developing a soil quality/health report card. Participatory on-farm research and education for agricultural sustainability. p. 223–224. Univ. of Illinois, Urbana-Champaign, IL.

Larson, W.E., and F.J. Pierce. 1991. Conservation and enhancement of soil quality. In Evaluation for sustainable land management in the developing world. Vol. 2. Technical papers. IBSRAM Proceedings No. 12 (2). Board for SOil Res. and Manage. Bangkok, Thailand.

Papendick, R.I., and J.F. Parr. 1992. Soil quality—the key to a sustainable agriculture. Am. J. Altern. Agric. 7:2–3.

Porter, P.A. 1991. Soil biological health: What do Wisconsin farmers think? NPM Field Notes, Vol. 2 (8):3. Center for Integrated Agric. Systems, Univ. of Wisconsin, Madison, WI.

Porter, P.A., M.J. Garlynd, A.V. Kurakov, and R.F. Harris. 1992. Wisconsin farmer perceptions of soil health. p. 265. *In* Agronomy abstracts. ASA, Madison, WI.

11 Biologically Active Soil Organics: A Case of Double Identity

J.A.E. Molina and H.H. Cheng

University of Minnesota
St. Paul, Minnesota

B. Nicolardot and R. Chaussod

Institut National de la Recherche Agronomique (INRA)
Dijon, France

Sabine Houot

Institut National de la Recherche Agronomique (INRA)
Thiverval-Grigon, France

One of the most sensitive, yet least defined, indicators of soil quality is its bioactivity. This indicator is often characterized, albeit inadequately, by the soil organic matter, or soil organic C contents. A more appropriate indicator, however, would be the biologically active components of the soil organic matter. These components are constitutents of the microbial biomass and a variety of soil organics serving as energy sources for the microbes. Biologically active soil organics can be defined by two methodological approaches: (i) the *experimental approach*, which considers the active organics as those obtained by a chemical fractionation procedure, and (ii) the *mathematical abstraction*, which represents active organics as time-dependent variables in a system of differential equations depicting soil C and N transformations. Although many methods have been proposed to define biologically active soil organic components either experimentally or methematically, there have been few reports on attempts to identify by chemical analysis the organic pools defined by the modelers, or to characterize by modeling the dynamic dimension of analytically determined organic fractions. Analytical chemists and computer modelers have not collaborated extensively in this area.

Only when this duality is bridged, will it be possible to measure how soil quality is affected by soil bioactivity and to predict how soil quality might be influenced by climatic, soil, and management changes. In this chapter,

[1]Contribution of the University of Minnesota. This work was supported by a grant from the Organization for Economic Co-Operation and Development (OECD).

a scheme is suggested as a possible approach to identify soil organic fractions corresponding to the dynamic pools of a model.

DYNAMIC ORGANIC POOLS

The modeling of C and N transformations in soil was pioneered by Jenny (1941) and Henin and Dupuis (1945) (Table 11-1). They were also the few who worked with models that were totally experimentally defined, because they considered only one dynamic pool: total organic C or N. Stanford and Smith (1972) pointed out that only a fraction of the total soil organic matter is active and introduced the concept of potentially mineralizable N (No). The meaning of No has been extensively debated by chemists (Keeney, 1982; Juma & Paul, 1984; Gianello & Bremner, 1986). Early modelers were limited by the necessity to work out the mathematics with paper and pencil. In 1973, Beek and Frissel used a computer to obtain a numerical solution to systems of differential equations, thus allowing modelers to work with a variety of dynamic pools.

Modeler's dynamic pools are defined by their position in the network of soil C and N flows, and their stability. In Table 11-1, the dynamic pools are categorized along a stability scale represented by their decay rate constants. Active pools, which disappear rapidly if not resupplied with an influx of C and N, have a low rate constant; the stable forms are found at the other end of the scale. Table 11-1 may thus be viewed as a representation of the modelers' communal wisdom in regard to numbers and stability of dynamic pools. It is best visualized by summing up the number of pools per category of rate constants and thus obtain a histogram of dynamic pool stability (Fig. 11-1). The striking feature of this histogram is its continuity, indicating that perhaps soil organics are synthesized through the gradual rearrangement of plant residues toward increasingly less decomposable forms. Some models are indeed built on such a hypothesis (Bosatta & Agren, 1985). Yet, experimental determinations of the half-life of soil organic fractions would indicate there are gaps in the stability spectrum (Hsieh, 1992) associated with soil textural properties (Martel & Paul, 1974; Balesdent et al., 1987), as is assumed by most modelers.

DETERMINATION OF INITIAL LEVELS
OF DYNAMIC ORGANIC POOLS

Abstract models have to be justified or validated against experimental data. This is achieved with those dynamic pools that have unequivocal chemical counterparts: inorganic species (CO_2, NH_4, and NO_3), and total organic C and N. For a model with n pools, the system of n differential equations—the mathematical expression of a model—has therefore the following form:

Table 11-1. Organic pools used in models of soil C and N transformations.

References	Element considered		Organic pools†	Decay rate constant (d^{-1})								
	C	N		1.0	$1 \cdot 10^{-1}$	$1 \cdot 10^{-2}$	$1 \cdot 10^{-3}$	$1 \cdot 10^{-4}$	$1 \cdot 10^{-5}$	$1 \cdot 10^{-6}$	$1 \cdot 10^{-7}$	0.0
Jenny, 1941		X	Initial soil N				X					
Henin & Dupuis, 1945	X		Soil organic matter (SOM)					X				
Stanford & Smith, 1972		X	Potentially mineralizable N (No)			X						
Beek & Frissel, 1973	X	X	Microbial biomass	X								
			Humus									X
Hunt, 1977	X		Humic material						X			
Jenkinson & Rayner, 1977	X	X	Microbial biomass			X						
			Physically protected OM					X				
			Chemically protected OM					X				
Paul & Van Veen, 1978	X	X	Microbial biomass		X							
			Decomposable SOM not physically protected		X							
			Decomposable SOM physically protected				X					
			Recalcitrant SOM not physically protected						X			
			Recalcitrant SOM physically protected						X			
Frissel & Van Veen, 1981	X	X	Biomass developed on available compounds		X							
			Biomass developed on resistant compounds			X						
			N containing decomposable materials C + N‡									

(continued on next page)

Table 11-1. Continued.

References	Element considered		Organic pools†	Decay rate constant (d^{-1})																
	C	N		1.0		1.10^{-1}		1.10^{-2}		1.10^{-3}		1.10^{-4}		1.10^{-5}		1.10^{-6}		1.10^{-7}		0.0
Frissel & Van Veen, 1981 (continued)			Decomposable active materials C‡																	
			Resistant active materials C + N						X											
			Old OM C + N										X							
Paul & Juma, 1981	X	X	Microbial biomass				X													
			Metabolites C + N	X																
			Metabolites C	X																
			Active fraction						X											
			Stabilized C + N								X									
			Old C + N												X					
Molina et al., 1983	X	X	Pool I labile		X															
			Pool I resistant				X													
			Pool II labile		X															
			Pool II resistant						X											
Van Veen et al., 1985	X	X	Unprotected microbial biomass		X															
			Protected microbial biomass						X											
			Decomposable metabolites		X															
			Recalcitrant metabolites		X															
			Active protected SOM								X									
			Old resistant SOM														X			
Gilmour et al., 1985	X	X	Microbial biomass I						X											
			Microbial biomass II								X									
			SOM I								X									
			SOM II																	X
Parton et al., 1987	X	X	Active SOM						X											
			Slow SOM								X									
			Passive SOM												X					

Reference		Pool						
Jenkinson et al., 1987	X	Zymogenous microbial biomass		X				
		Autochtonous microbial biomass		X				
		Humidified OM					X	
		Inert OM						X
Buyanovsky et al., 1987	X	Aboveground microbial biomass		X				
		Underground microbial biomass		X				
		SOM				X		
Andren & Paustian, 1987	X	Active SOM	X					
		Stabilized SOM			X			
Van der Linden et al., 1987	X	Microbial biomass	X					
		Easily decomposable fraction	X					
		Recalcitrant fraction			X			
		Old SOM				X		
Bjarnason, 1989	X	Labile SOM	X					
		Stable SOM				X		
Verberne et al., 1990	X	Non protected microbial biomass	X					
		Protected microbial biomass		X				
		Non protected OM	X					
		Protected OM			X			
		SOM				X		
Almendinger, 1990	X	SOM	X			X		

† Plant residues and litter pools excluded.
‡ Decomposition of these pools follows Michaelis-Menten kinetics, instead of first-order kinetics.

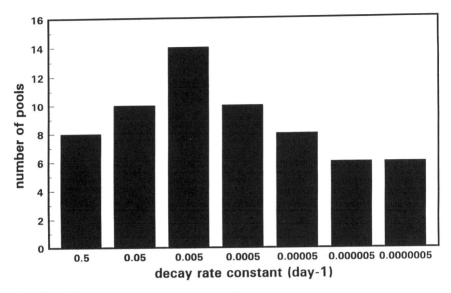

Fig. 11-1. Histogram of soil organic pool stability.

$$dY_i/dt = F_i(Y_i, Z_j, K)$$

$$dZ_j/dt = F_j(Y_i, Z_j, K)$$

where Y_i ($i = 1 \ldots m$) are the time-dependent dynamic pools, each with an experimentally defined equivalent (e.g., NH_4, NO_3, CO_2, and total organic C and N); Z_j ($j = m + 1 \ldots n$) are the time-dependent speculated organic pools, which have not yet been experimentally measured; and K is a vector constant, which includes the initial values for the Y and Z variables (Y_{oi}, Z_{oj}) and the parameters of biological transformations such as microbial efficiency factors, decay rate constants, and C/N ratios of the organic pools Z_j. The F functions, which express for each pool the rates of ongoing and outgoing flows, are a representation of the mechanistic concepts underlying the architecture of the network of C and N flows between the dynamic pools (Jansson & Persson, 1982; Molina et al., 1990).

The model is defined and the kinetics of Y_i and Z_j can be simulated when values have been assigned to the constants. Ideally, these values should be measured. This can be done only for Y_{oi} and some of the speculated parameters. The other constants—in particular the initial values for the organic pools Z_{oj}—are obtained by optimization against the experimental kinetics Y_i; although a legitimate procedure, the optimization step weakens the validation process, because it uses the information contained in Y_i to simultaneously validate and define the model (Barak et al., 1990). The lower the number of experimental kinetics Y_i with respect to the number of constants optimized (e.g., the number of speculated dynamic pools in a model), the easier it is to obtain a good fit between measured and simulated Y_i values; this is a situation not unlike the certainty to see n or less points touched by the graph of a polynomial of degree n. Under such circumstances, the model may provide a *good mimic* of the experimental data, but the probability that it correctly represents soil C and N transformations is low.

ESTABLISHING A DIALOGUE BETWEEN MODELERS AND ANALYTICAL CHEMISTS

Soil organic matter has been fractionated by countless classical techniques. Recently, novel technologies have offered the prospect of a renewed surge of activity (e.g., Cambardella & Elliott, 1992; Cheng et al., 1993; Schnitzer & Schulten, 1992). The challenge is to have the basis to establish a one-to-one correspondence between those dynamic organic pools found in a model, and the analytically determined soil organic fractions. To that effect, the following approach is proposed.

As a first step it behooves the modelers to work with a model that has been validated with as high a level of probability as possible. This requires drawing from one soil as much information from the determination of CO_2, inorganic N, and total organic C and N as techniques will allow—such as using tracer C and N and performing incubations with the same soil for different controllable conditions (e.g., different initial levels of NH_4 and sources of tracer C, different temperatures of incubation, etc.). The incubation should be carried long enough to ensure gradual dissemination of tracer C and N throughout the organic pools. Once a successful optimization has been obtained under such stringent conditions, the kinetics of organic pools—tracer and nontracer C and N—are simulated and assumed to be a true representation of changes with time of the soil organic fractions. The analytical chemist will then have the basis to develop appropriate extraction procedures. It is sufficient to search for those soil organic fractions that will provide measured kinetics similar to the simulated ones.

This procedure will test the hypothesis that there are soil organic fractions that can be experimentally identified, and accumulate C and N according to the flows and kinetics described by the model. Rejection of the hypothesis carries two risks: (i) the correct analytical procedure has not been tried; and (ii) the model is a misrepresentation of the actual C and N flow. In the first case the probability of reaching the wrong conclusion is lowered with a higher number of separation techniques tried. The second risk can be minimized by the use of models previously validated under the stringent conditions described above.

SUMMARY

Biologically active soil organic C and N are important indicators of soil quality. They can be categorized by dynamic pools of varying rate constants in simulation models. They are also obtained by chemical extractions of soil organic fractions. A method is proposed to identify soil organic fractions corresponding to the dynamic pools of models.

REFERENCES

Almendinger, J.C. 1990. The decline of soil organic matter, total nitrogen and available water capacity following the late-holocene establishment of jack pine on sandy Mollisols in north central Minnesota (USA). Soil Biol. Biochem. 150:680–694.

Andren, O., and K. Paustian. 1987. Barley straw decomposition in the field: A comparison of models. Ecology 68:1190–1200.

Balesdent, J., A. Mariotti, and B. Guillet. 1987. Natural abundance as a tracer for studies of soil organic matter dynamics. Soil Biol. Biochem. 19:25–30.

Barak, P., J.A.E. Molina, A. Hadas, and C.E. Clapp. 1990. Optimization of an ecological model with the Marquardt algorithm. Ecol. Model. 51:251–263.

Beek, J., and M.J. Frissel. 1973. Simulation of nitrogen behaviour in soils. Pudoc, Wageningen, the Netherlands.

Bjarnason, S. 1989. The long-term soil fertility experiments in southern Sweden: III. Soil carbon and nitrogen dynamics. Acta Agric. Scand. 39:361–372.

Bosatta, E., and G.I. Agren. 1985. Theoretical analysis of decompositoin of heterogeneous substrates. Soil Biol. Biochem. 17:601–610.

Buyanovsky, G.A., C.L. Kucera, and G.H. Wagner. 1987. Comparative analyses of carbon dynamics in native and cultivated ecosystems. Ecology 68:2023–2031.

Cambardella, C.A., and E.T. Elliott. 1992. Particulate soil organic changes across a grassland cultivation sequence. Soil Sci. Soc. Am. J. 56:777–783.

Cheng, H.H., J. Gan, W.C. Koskinen, and L.J. Jarvis. 1993. Potential of the supercritical fluid extraction technique for characterizing organic–inorganic interactions in soils. Proc. Int. Soc. Soil Sci. Working Group MO: Impact of Interactions of Inorganic, Organic, and Microbiological Soil Components on Environmental Quality, Edmonton, Canada. 12–15 Aug. 1992.

Frissel, M.J., and J.A. Van Veen. 1981. Simulation model for nitrogen immobilization and mineralization. p. 359–381. In I.K. Iskandar (ed.) Modeling wastewater renovation—land treatment. John Wiley & Sons, New York.

Gianello, C., and J.M. Bremner. 1986. Comparison of chemical methods of assessing potentially available organic nitrogen in soil. Commn. Soil Sci. Plant Anal. 17:215–236.

Gilmour, J.T., M.D. Clark, and G.C. Sigua. 1985. Estimating net nitrogen mineralization from carbon dioxide evolution. Soil Sci. Soc. Am. J. 49:1398–1402.

Henin, S., and M. Dupuis. 1945. Essai de bilan de la matiere organique du sol. Ann. Agron. 15:17–29.

Hsieh, Y.P. 1992. Pool size and mean age of stable soil organic carbon in cropland. Soil Sci. Soc. Am. J. 56:460–464.

Hunt, H.W. 1977. A simulation model for decomposition in grasslands. Ecology 58:469–484.

Jansson, S.L., and J. Persson. 1982. Mineralization and immobilization of soil nitrogen. p. 229–252. In F.J. Stevenson (ed.) Nitrogen in agricultural soils. Agron. Monogr. 22. ASA, CSSA, and SSSA, Madison, WI.

Jenkinson, D.S., P.B.S. Hart, J.H. Rayner, and L.C. Parry. 1987. Modelling the turnover of organic matter in long-term experiments at Rothamsted. Intecol Bulletin 15:1–8.

Jenkinson, D.S., and J.H. Rayner. 1977. The turnover of soil organic matter in some of the Rothamsted classical experiments. Soil Sci. 123:298–305.

Jenny, H. 1941. Factors of soil formation—system of quantitative pedology. McGraw-Hill, New York.

Juma, N.G., and E.A. Paul. 1984. Mineralizable soil nitrogen: Amounts and extractability ratios. Soil Sci. Soc. Am. J. 48:76–80.

Keeney, D.R. 1982. Nitrogen—availability indices. p. 711–730. In A.L. Page (ed.) Methods of soil analysis. Part 2. 2nd ed. Agron. Monogr. 9. ASA and SSSA, Madison, WI.

Martel, Y.A., and E.A. Paul. 1974. The use of radiocarbon dating of organic matter in the study of soil genesis. Soil Sci. Soc. Am. Proc. 38:501–506.

Molina, J.A.E., C.E. Clapp, M.J. Shaffer, F.W. Chichester, and W.E. Larson. 1983. NCSOIL, a model of nitrogen and carbon transformations in soil: Description, calibration and behavior. Soil Sci. Soc. Am. J. 47:85–91.

Molina, J.A.E., A. Hadas, and C.E. Clapp. 1990. Computer simulation of nitrogen turnover in soil and priming effect. Soil Biol. Biochem. 22:349–353.

Parton, W.J., D.S. Schimel, C.V. Cole, and D.S. Ojima. 1987. Analysis of factors controlling soil organic matter levels in great plains grasslands. Soil Sci. Soc. Am. J. 51:1173–1179.

Paul, E.A., and N.G. Juma. 1981. Mineralization and immobilization of soil nitrogen by microorganisms. *In* F.E. Clark and T. Rosswall (ed.) Terrestrial nitrogen cycles—processes, ecosystem strategies and management impacts. Ecol. Bull. 33:179–199.

Paul, E.A., and J.A. Van Veen. 1978. The use of tracers to determine the dynamic nature of organic matter. Trans. 11th Int. Congr. Soil Sci. 3:61–102.

Schnitzer, M., and H.R. Schulten. 1992. The analysis of soil organic matter by pyrolysis-filed ionization mass spectrometry. Soil Sci. Soc. Am. J. 56:1811–1817.

Stanford, G., and S.J. Smith. 1972. Nitrogen mineralization potentials of soils. Soil Sci. Soc. Am. Proc. 36:465–472.

Van der Linden, A.M.A., J.A. Van Veen, and M.J. Frissel. 1987. Modelling soil organic matter levels after long-term applications of crop residues, and farmyard and green manures. Plant Soil 101:21–28.

Van Veen, J.A., J.N. Ladd, and M. Amato. 1985. Turnover of carbon and nitrogen through the microbial biomass in a sandy loam and a clay soil incubated with [14C(U)]glucose and [15N] $(NH_4)_2SO_4$ under different moisture regimes. Soil Biol. Biochem. 17:747–756.

Verberne, E.L.J., J. Hassink, P. DeWilligen, J.J.R. Groot, and J.A. Van Veen. 1990. Modelling organic matter dynamics in different soils. Neth. J. Agric. Sci. 38:221–238.

12 Terrestrial Carbon Pools: Preliminary Data from the Corn Belt and Great Plains Regions[1]

Edward T. Elliott, Ingrid C. Burke, Christopher A. Monz, and Serita D. Frey

Colorado State University
Fort Collins, Colorado

Keith H. Paustian and Harold P. Collins

Kellogg Biological Station
Michigan State University
Hickory Corners, Michigan

Eldor A. Paul

Michigan State University
East Lansing, Michigan

C. Vernon Cole

USDA-ARS
Colorado State University
Fort Collins, Colorado

Robert L. Blevins and Wilbur W. Frye

University of Kentucky
Lexington, Kentucky

Drew J. Lyon

University of Nebraska
Scotts Bluff, Nebraska

Ardell D. Halvorson

USDA-ARS
Akron, Colorado

David R. Huggins

Southwest Experiment Station
Lamberton, Minnesota

Ronald F. Turco

Purdue University
West Lafayette, Indiana

Michael V. Hickman

USDA-ARS
Purdue University
West Lafayette, Indiana

Soil organic matter is recognized as an important component of soil quality (Granatstein & Bezdicek, 1992; Arshad & Coen, 1992). In mineral soils, many properties associated with soil quality, including nutrient mineralization, aggregate stability, trafficability, and favorable water relations are related to the soil organic matter content. Past considerations of soil organic matter and how it is affected by management practices have largely reflected the

[1] This research is supported by the USEPA (Project AERL-9101) and the USDA-ARS.

importance of organic matter to soil fertility and crop production. More recently, interest in soil organic matter and its relationship to agricultural management has developed with respect to its role in the worldwide C budget and worldwide climate change, another important quality of soil.

The conversion of native ecosystems to agriculture usually results in a loss of soil organic matter. Such losses are well-documented for the Great Plains and Corn Belt regions of North America (Jenny, 1941; Haas & Evans, 1957; Campbell, 1978). Worldwide, soil C losses associated with agricultural expansion and intensification since the mid-1800s have contributed significantly to increased atmospheric CO_2 (Post et al., 1990; Wilson, 1978). There is growing interest in reversing this trend through changes in agricultural practices to make soil a net sink for C.

The research described in this chapter was initiated to assess the potential for C sequestration in agricultural soils of the Great Plains and Corn Belt regions (Barnwell et al., 1992). However, understanding the factors governing changes in soil organic matter pool sizes (e.g., climatic, edaphic, and management factors) is also crucial for assessment of other essential aspects of soil quality and its response to management.

Data from long-term field studies has provided much information about changes in soil C under various management regimes, but our knowledge base is fragmented. Most of what we know concerning soil organic matter (SOM) dynamics has been obtained by studying SOM losses. We understand less about controls on SOM accumulation and how they vary across soil types, climatic regions, and management regimes. Such information is crucial for estimating the potential for C sequestration in agricultural soils of the Great Plains and Corn Belt regions.

Our approach to assessing potential C sequestration in agroecosystems is based on relating key driving variables to ecosystem processes and properties. Driving variable control can explain soil properties and their geographic distribution (Jenny, 1941; Elliott et al., 1993). This framework acknowledges that soils are one of many interrelated ecosysem components.

Understanding soil C dynamics and assessing potential C sequestration in soil requires investigation of the functioning of the whole ecosystem with its inherent interactions and feedbacks. Relationships between soil C management, climate, and soil type cannot be adequately resolved using empirical statistical approaches because of these interactions and feedbacks. The best way to accomplish this integration is through the use of simulation models (e.g., Parton et al., 1987; Paustian et al., 1992; Jenkinson et al., 1987).

Agroecosystem models must be validated over a wide range of management practices, climates, soils, and time scales to be generally applicable. Fortunately, there are appropriate data available from long-term field experiments (Elliott & Cole, 1989). These experiments are located on research sites of the USDA-ARS, state experiment stations, Agriculture Canada, and other countries. They constitute the largest and longest-term set of ecological experiments currently available. However, this information is not organized into comprehensive data bases and essential information has not been collected at every site.

The objective of this research is to develop reliable data sets to assess the potential for C sequestration in agroecosystems of the Great Plains and Corn Belt. To supplement existing information, we are sampling a number of long-term field experiments to obtain a uniform characterization of soil organic matter and organic matter inputs relating to specific active fraction pools (particulate organic matter, $CHCl_3$, labile C, and mineralized C) useful for parameterizing and validating SOM models. This chapter represents the spring sampling of six sites within the current network of 41 sites (Fig. 12-1), three from the Corn Belt and three from the Great Plains.

MATERIALS AND METHODS

Sampling

All sites (Table 12-1) were sampled in the spring of 1992 prior to cultural operations. Six 5.6-cm cores were collected and composited from each field replicate using a truck-mounted hydraulic soil exploration probe. Each soil core was divided into four depth increments; 0 to 20, 20 to 25, 25 to 50, and 50 to 100 cm. The 0 to 20 and 20 to 25 cm increments allow us to adjust for potential differences in bulk density in different management practices so that we can express our data based on an equivalent soil weight as well as an equivalent soil depth. For no-till and grassland treatments, six additional 5.6 by 2.5 cm cores were collected and composited from the surface 2.5 cm of mineral soil. In addition, surface residues from three 0.25 m^2 areas were collected at random within each no-till field replicate.

Immediately after collection, soil samples were put on ice for transport to the laboratory. Surface (0–20 cm) samples used in biological analyses were stored at 4 °C for no more than 24 h and sieved moist to pass a 2-mm screen prior to incubation. Remaining samples were air-dried, crushed, then sieved after removal of large plant fragments. Crop residues, roots, and rock fragments within each sample were rmeoved, dried, and weighed. All data reported in this chapter are reported to a 20-cm depth.

Soil Analysis

Total C was determined by dry combustion of duplicate subsamples from each depth for each treatment replicate on a Carlo Erba CHN analyzer Model 1104 (Carlo Erba Instruments, Milano, Italy). Soil texture and particulate organic matter (POM-C) were determined by a modified version of the Cambardella and Elliott (1992) method. Briefly, POM was isolated by dispersing 30 g of soil in 5% sodium hexametaphosphate and passing the dispersed sample through a 53-μm sieve, which retains the POM fraction plus sand and allows the passage of mineral-associated SOM. The sand plus POM fraction was dried (50 °C), finely ground, and subsampled for total organic C. The soil slurry passing through the sieve was transferred to a 1-L sedimentation cylinder, and the silt and clay content determined using the hydrometer

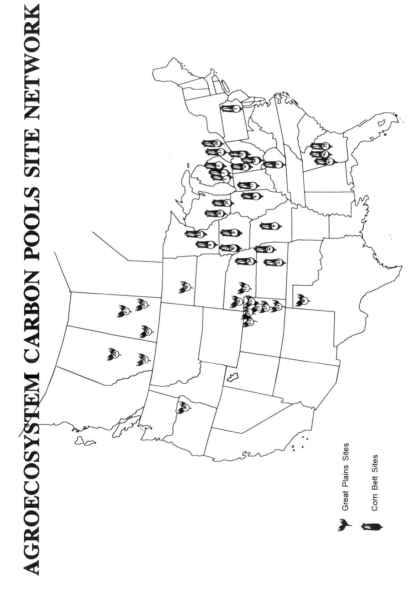

Fig. 12–1. The C Pools and Dynamics site network is made up of collaborating sites represented by cooperating scientists. The current 41 sites are distributed across the Corn Belt and Great Plains with some sites located in other parts of the country. Data included in this chapter are from samples taken in the spring of 1992 at three sites in the Corn Belt (5, 8, and 11) and three sites in the Great Plains (15, 16, and 24).

Table 12-1. Background and treatments sampled for this study.

Site (soil series) [soil classification] key citation	Duration of experiment (yr)	Rotation	Tillage	Reps	Fertilization (kg N ha^{-1})	No. plots
Lamberton, MN: (Normania loam) [Aquic Haplustoll] Nelson & MacGregor, 1973	32	Continuous corn	Moldboard	4	16, 106, 195	12
Lexington, KY: (Maury silt loam) [Typic Paleudalf] Blevins et al., 1983	22	Continuous corn	Moldboard No-till	4	0, 84, 164	24
W. Lafayette, IN: (Chalmers silty clay loam) [Typic Haplustoll] and (Raub silt loam) [Aquic Haplaquoll] Schreiber et al., 1987	11	Continuous corn Corn–soybean	Moldboard No-till	4	92	16
Sidney, NE: (Duroc loam) [Pachic Haplustoll] Fenster & Peterson, 1979	22	Wheat–fallow	Moldboard Stubble mulch No-till Native grassland	3	None	12

(continued on next page)

Table 12-1. Continued.

Site (soil series) [soil classification] key citation	Duration of experiment	Rotation	Tillage	Reps	Fertilization	No. plots
Akron, CO: (Weld silt loam) [Aridic Paleustoll] Smika, 1990	25	Wheat–fallow	Stubble mulch No-till Native grassland	3	50 every 2 yr	9
Pawnee Grassland 1: (Ascalon fine sandy loam) [Aridic Argiustoll] Coffin et al. (unpubl.)	~50†	Moldboard (when cultivated)	Successional go-back Native grassland	3	None	6
Pawnee Grassland 2: (Ascalon fine sandy loam) [Aridic Argiustoll] Coffin et al. (unpubl.)	~50	Moldboard (when cultivated)	Successional go-back Native grassland	3	None	6
Pawnee Grassland 3: (Ascalon fine sandy clay loam) [Aridic Argiustoll] Coffin et al. (unpubl.)	~50	Moldboard (when cultivated)	Successional go-back Native grassland	3	None	6

† Years since abandonment.

method. Bulk density was determined from the number of cores collected and volume of the core that each depth increment represented.

Moist soil subsamples from the 0 to 20-cm depth increment were used to determine mineralized C and $CHCl_3$ labile C. Field capacity water content for each soil sample was determined by a volumetric soil water method. These water contents were used for subsequent soil incubations. Briefly, dried and sieved soils were packed into 50 cm^3 tubes. Enough water was added to wet approximately half of the soil in the tube. The soil in the columns was allowed to equilibrate with the water for 18 h, after which the wet soil was removed and moisture content determined. Even though matric potential measurements were available for most soils sampled, we decided to use the above method to assure uniformity of moisture content across the many soil types.

Mineralized soil C was determined by incubating duplicate 25-g soil samples in 160-mL bottles at 25 °C for 91 d. Headspace CO_2 was measured using a Beckman Model 865 infrared gas analyzer (Beckman Instruments, Fullerton, CA) at 2, 4, 7, 10, and then 14 d intervals to 91 d. Following analysis, each sample was returned to ambient CO_2 by de-gassing with compressed air.

Chloroform labile C (i.e., released microbial biomass C) was estimated by the chloroform fumigation–extraction method as described by Vance et al. (1987) and Tate et al. (1988), except that 12 g of moist soil was used rather than the usual 25 g dry weight equivalent. Prior to fumigation, soil samples were preincubated at field capacity for 7 d (25 °C) to normalize vagarities of weather relative to sampling data across sites. Soluble organic C was extracted from duplicate fumigated and control samples using 0.5 M K_2SO_4 and oxidized in a sulfuric–phosphoric acid dichromate mixture (Snyder & Trofymow, 1984). Carbon dioxide liberated during digestion was trapped in 0.1 M NaOH and titrated with HCl (Anderson, 1982).

RESULTS

Corn Belt Sites (Table 12–2)

Lamberton, MN

Total soil C to the 20-cm depth was not significantly different among fertilizer treatment levels. Active fraction components were similar across treatments, although there was a weak trend toward increased soil C pools with increased N fertilizer rates. There was significantly less POM in the lowest N application treatment and, unexpectedly, less $CHCl_3$ labile C in the highest N treatment.

Lexington, KY

Soil C was greater with increasing fertilizer additions in the conventionally tilled (CT) treatment, but was not different among fertility levels in the

Table 12-2. Amounts of total C and active fraction C pools for selected Corn Belt sites.

Site treatment	Total	POM†	CHCl₃ labile	Mineralized‡	Sand	Silt	Clay	Bulk density
	—		Mg C ha^{-1} (to 20 cm)	—	—	g g^{-1}	—	Mg m^{-3}
				Lamberton, MN				
CT-16§	56a*	3.0a	0.34a	1.4a	0.37a	0.33a	0.30a	1.38a
CT-106	61a	4.6b	0.41b	1.5a	0.33a	0.34a	0.33a	1.35a
CT-195	62a	4.4b	0.27b	1.6a	0.36a	0.34a	0.30a	1.33a
				Lexington, KY				
CT-0	30a	3.6a	0.29a	2.1a	0.07a	0.63a	0.30b	1.27a
CT-84	34ab	4.7a	0.30a	2.5ab	0.07a	0.63a	0.30b	1.27a
CT-164	40b	6.3b	0.23a	3.0c	0.08a	0.64ab	0.29b	1.29a
NT-0	37b	4.3a	0.42b	2.4ab	0.07a	0.64ab	0.29b	1.32a
NT-84	40b	6.3b	0.23a	2.6b	0.08a	0.66b	0.27a	1.28a
NT-164	37b	7.0b	0.30a	3.3c	0.08a	0.64ab	0.28ab	1.21a
				W. Lafayette, IN				
CT-CS	60a	5.3a	0.28a	2.1a	0.15a	0.51a	0.34a	1.20a
CT-CC	77a	6.4a	0.35a	2.7bc	0.14a	0.49a	0.37a	1.25ab
NT-CS	73a	7.6a	0.26a	2.4ab	0.15a	0.49a	0.34a	1.23ab
NT-CC	65a	10.1a	0.32a	3.0c	0.15a	0.53a	0.32a	1.31b
		Conventional tillage—continuous corn across sites						
MN	61b	4.6a	0.41a	1.5a	0.33c	0.34a	0.33a	1.35a
KY	34a	4.7a	0.30a	2.5b	0.07a	0.63c	0.30a	1.27a
IN	77c	6.4b	0.35a	2.7b	0.14b	0.49b	0.37b	1.25a

* For comparisons across tillage or fertility treatments within each site, values followed by the same letter are not different at $P = 0.05$ using LSD mean separation tests.
† POM = particulate organic matter.
‡ Cumulative respiration at 91 d.
§ CT = conventional tillage; NT = no-tillage; CC = continuous corn rotation; CS = corn-soybean rotation (all crops at Lamberton and Lexington are continuous corn); numbers following treatment indicate fertilizer additions in kg N ha^{-1} yr^1.

no-till (NT) treatment. No-till organic matter levels were higher than CT at the lowest fertility level, but not at the higher two levels. These trends were essentially repeated for the POM fraction. Chloroform extractable C was not different among treatments, except at the lowest fertilizer level where NT was greater than CT. Mineralized C was greater at the highest compared with lowest fertility levels within each cultivation treatment.

West Lafayette, IN

Within-treatment variability was high for this site, perhaps due to the irregular occurrence of two different soil series (see Table 12-1) across the plots. Total C and active fraction components were not significantly differ-net for tillage method or crop rotation, except for mineralized C. Mineralized C was significantly higher in continuous corn (*Zea mays* L.) than in corn-soybean [*Glycine max* (L.) Merr.] rotations and there was a trend for it to increase under no-till compared with conventional tillage.

Cross-Site Comparison of Continuous Corn with Conventional Tillage

Cross-site comparisons were made for the common treatment of continuous corn, moldboard plow tillage, and N fertilizer rates of ~80 to 100 kg ha^{-1} yr^{-1}. The two prairie-derived soils (Lamberton and W. Lafayette) had more total soil C than the forest-derived soil (Lexington). Soil textures differed primarily in sand and silt content, with the Kentucky and Indiana sites having the finest textures. Interestingly, the Lamberton site showed a much lower activity for mineralized C and the relative size of all measured active C fractions compared to total carbon was lower than for the other two sites.

Great Plains Sites (Table 12–3)

Sidney, NE

Bulk density was greater in the bare fallow and NT than in stubble mulch and native grassland treatments. The native grassland and NT treatments had greater amounts of total soil C than bare fallow and stubble mulch treatments. There was more POM-C in the native grassland treatment than the other treatments and NT had more POM than stubble mulch or bare fallow. Chloroform labile C was greater in the NT treatment than bare fallow. Mineralized C increased with decreasing soil disturbance due to tillage.

Akron, CO

Sand content of the soil was greater in the native grassland than stubble mulch or NT treatments, whereas the bulk density was lowest in the native grassland treatment. The stubble mulch and NT treatments were not different from each other and were less than the native grassland treatment for total C and all active fraction soil C pools.

Abandoned and Native Grassland Pawnee Grassland Treatments

No differences were observed between abandoned and native treatments for Sites 1 and 2, except at Site 1 where mineralized C in the abandoned treatment was greater than that in the native grassland treatment. At Site 3, the total soil C and all active fractions were greater in the native than the abandoned treatments.

Comparison of Native Grassland Treatments across Sites

Sand related directly to the bulk density of the undisturbed soils. Patterns within the three Pawnee Grassland sites showed that increasing clay content positively correlated with all measured soil C pools. The Sidney site had the most precipitation (data not shown) and the highest levels of total C and POM-C. Pawnee Grassland Site 3 had the highest clay content and microbial biomass levels.

Table 12-3. Amounts of total C and active fraction C pools for selected Great Plains sites.

Site treatment	Total	POM†	CHCl₃ labile	Mineralized‡	Sand	Silt	Clay	Bulk density
	— Mg C ha⁻¹ (to 20 cm) —				— g g⁻¹ —			Mg m⁻³
				Sidney, NE				
BF§	28a*	3.5a	3.37ab	1.4a	0.41b	0.32a	0.26a	1.13a
SM	28a	4.2a	0.33a	1.5a	0.35a	0.37a	0.28a	0.92a
NT	36b	5.1b	0.58b	2.2b	0.33a	0.38a	0.29a	1.10b
NG	38b	11.1c	5.52ab	2.7c	0.35a	0.42a	0.23a	0.94a
				Akron, CO				
SM	17a	2.5a	0.32a	1.5a	0.34a	0.42a	0.27b	1.16ab
NT	15a	2.5a	0.33a	1.5a	0.36a	0.39a	0.25ab	1.23b
NG	25b	8.1b	0.60b	2.3b	0.44b	0.35a	0.20a	1.01a
				Successional grasslands				
Pawnee Grassland 1								
AB	18NS	3.4NS	0.64NS	2.4*	0.71NS	0.13NS	0.17NS	1.33NS
NG	20	3.6	0.65	1.5	0.72	0.13	0.15	1.36
Pawnee Grassland 2								
AB	22NS	5.4NS	1.08NS	2.7NS	0.60NS	0.25NS	0.15NS	1.34NS
NG	24	6.2	0.87	2.7	0.52	0.30	0.19	1.23
Pawnee Grassland 3								
AB	24*	5.3*	0.94*	2.6*	0.53NS	0.15NS	0.32NS	1.29NS
NG	30	8.1	1.11	3.3	0.54	0.17	0.30	1.24
				Native grassland comparison				
Sidney	38c	11.1c	0.52a	2.7c	0.35a	0.42c	0.23ab	0.94a
Akron	25ab	8.1b	0.60a	2.3b	0.44ab	0.35c	0.20a	1.01a
Pawnee Grassland 1	20a	3.6a	0.65a	1.5a	0.72c	0.12a	0.15a	1.36b
Pawnee Grassland 2	24a	6.2b	0.87b	2.7cd	0.52b	0.30bc	0.19a	1.23b
Pawnee Grassland 3	30b	8.1b	1.11c	3.3d	0.54b	0.17ab	0.29b	1.24b

* Significant at $p = 0.05$. Values followed by the same letter are not different at $p = 0.05$ using LSD mean separation tests; NS = not significant.
† POM = particulate organic matter.
‡ Cumulative respiration at 91 d.
§ BF = bare fallow; SM = stubble mulch; NT = no-tillage; NG = native grassland; AB = abandoned.

DISCUSSION

The largest differences among treatments were observed between the native and cultivated systems. Relative differences across sites within native grasslands, for the Great Plains sites, and within conventional tillage-continuous corn, for the Corn Belt sites, were greater than most differences observed within sites across management treatments. The more mesic Corn Belt sites had larger amounts of soil C than the Great Plains sites due to the greater primary production, hence higher C inputs leading to higher total C levels. The Corn Belt soils developed under tallgrass prairie or deciduous forest (Lexington) and the Great Plains soils developed under short or midgrass prairie. Interestingly, the active fraction pools were similar between the two regions, suggesting that potentially a higher proportion of total soil C was biologically active in the Great Plains than Corn Belt region.

In many cases relative differences among treatments was greater for active fraction pools than total soil C, but the number of treatments that were statistically different from each other within a site were no greater. This trend indicates that although the active fraction pools may be more sensitive to treatment influences, they are also more variable.

Corn Belt

Carbon mineralization and POM-C appear to be the most sensitive indicators of management effects on SOM within this region. The greater POM-C and mineralized C levels for NT, in particular, suggests increased potential for C availability to soil heterotrophs. Average organic matter levels to a depth of 20 cm were generally greater with reduced tillage and increased N application, but to varying degrees at different sites. Crop rotation effects on soil C pools, represented at the Indiana site, were clearly differentiated only for mineralized C, but variability in C pools across field replicates was much higher at W. Lafayette than the other two sites.

Although having a total soil C level intermediate between the other two sites, Lamberton, MN, exhibited lower C mineralization rates and smaller active C pools relative to total C. The continuous corn–conventional tillage management practice with 80 to 100 kg N ha^{-1} added every year yielded similar results for POM and biomass pools, but C mineralization rates underscored the strong differences in the amount of labile C in the Lamberton soil compared with the other two sites. In contrast to the cross-site comparison within the Great Plains, texture was not closely associated with soil C content. Lexington had the highest silt plus clay content but the lowest soil C, emphasizing the overwhelming influence of soil genesis on soil C levels.

Great Plains

Total soil C levels at the Sidney site were comparable in the native grassland and NT treatments, even 20 yr after being plowed out of grass. Tilled soils (stubble mulch and bare fallow) showed large losses of C. At Akron, where treatments were established in previously cultivated soil, the trend was different; the stubble mulch and NT treatments were similar in all respects and much lower than the native site. Precipitation at Akron is less than at Sidney, but Akron soils have been cultivated longer. Akron treatments (except native grassland) were initiated in degraded soils (~ 60 yr of cultivation), whereas Sidney treatments were begun on previously uncultivated grassland soils. A number of factors vary simultaneously between the two sites, so it is not possible to unambiguously explain observed differences. Data from additional sites will be needed to determine how management affects soil organic matter levels across the central USA.

Soils from the Pawnee Grassland abandoned from cultivation for approximately 50 yr have nearly recovered their original total soil C levels. At Site 2, which had the sandiest soil, the mineralized C in the abandoned treatment surpassed the native site. The native grassland treatment at Site 3 maintained higher levels of mineralized C compared with abandoned. This soil had the highest clay content.

Data presented in this article has demonstrated that management can control amounts of C in soils differentially under different soil–climate-management combinations. Further analysis of the wider range of network sites will assist understanding the extent to which increases in atmospheric CO_2 can be ameliorated by increases in total soil C as a result of management. Although we accept the many roles that soil organic matter plays in determining the quality of soil, it also influences the quality of the global enviornment as a sink for atmospheric CO_2.

SUMMARY

Soil organic matter is recognized as an important component of soil quality through its influence on soil physical properties and the cycling of nutrients. Interest in soil organic matter has expanded to include its role in the worldwide C budget and climate change. The research described here was initiated to help assess the potential for C sequestration in agricultural soils of the Great Plains and Corn Belt regions to ameliorate the buildup of atmospheric CO_2. In addition, a better understanding of what factors govern soil organic matter pool sizes is relevant to the assessment of soil quality and its response to management.

For the Corn Belt sites, C mineralization and POM-C were the most sensitive indicators of management effects on soil organic matter. No-till generally showed increased amounts of POM and mineralized C in the top 20 cm, suggesting increases in labile C fractions compared with conventional tillage. Mean organic matter levels to a depth of 20 cm generally increased with reduced tillage and higher N application, but to varying degrees at different sites.

Within the Great Plains sites, tilled soils (stubble mulch and bare fallow) showed large losses of C. Where NT was initiated on previously uncultivated soils at Sidney, NE, total soil C levels were comparable to the native grassland, even after 20 yr. No-till on previously degraded soil (~ 60 yr of cultivation) at Akron had C levels similar to stubble mulch and much lower than the native grassland. Carbon levels in two of three successional grassland sites had recovered to near those in the native systems after 50 yr. As with the Corn Belt sites, POM and mineralized C values showed the greatest differences among tillage treatments.

REFERENCES

Anderson, J.P.E. 1982. Soil respiration. p. 831–871. *In* A.L. Page (ed.) Methods of soil analysis. Part 2. 2nd ed. Agron. Monogr. 9. ASA and SSSA, Madison, WI.

Arshad, M.A., and G.M. Coen. 1992. Characterization of soil quality: Physical and chemical criteria. Am. J. Altern. Agric. 7:25–32.

Barnwell, T.O., R.B. Jackson, E.T. Elliott, I.C. Burke, C.V. Cole, K. Paustian, E.A. Paul, A.S. Donigian, A.S. Patwardhan, A. Rowell, and K. Weinrich. 1992. An approach to assessment of management impacts on agricultural soil carbon. Water Air Soil Pollut. 64:423–435.

Blevins, R.L., G.W. Thomas, M.S. Smith, W.W. Frye, and P.L. Cornelius. 1983. Changes in soil properties after 10 years continuous non-tilled and conventionally tilled corn. Soil Tillage Res. 3:135–146.

Cambardella, C.A., and E.T. Elliott. 1992. Particulate soil organic-matter changes across a grassland cultivation sequence. Soil Soc. Sci. Am. J. 56:777–783.

Campbell, C.A. 1978. Soil organic carbon, nitrogen and fertility. p. 173–271. In M. Schnitzer and S.U. Khan (ed.) Soil organic matter. Dev. in Soil Sci. 8. Elsevier, New York.

Elliott, E.T., and C.V. Cole. 1989. A perspective on agroecosystem science. Ecology 70:1597–1602.

Elliott, E.T., C.V. Cole, and C.A. Cambardella. 1993. Modification of ecosystem processes by management and the mediation of soil organic matter dynamics. p. 257–268. In K. Mulongoy and R. Mercx (ed.) The dynamics of soil organic matter in relation to the sustainability of tropical agriculture. John Wiley & Sons, New York.

Fenster, C.R., and G.A. Peterson. 1979. Effects of no-tillage fallow compared to conventional tillage in a wheat fallow system. Nebraska Agric. Exp. Stn. Res. Bull. 289.

Granatstein, D., and D.F. Bezdicek. 1992. The need for a soil quality index: Local and regional perspectives. Am. J. Altern. Agric. 7:12–16.

Haas, H.J., and C.E. Evans. 1957. Nitrogen and carbon changes in Great Plains soils as influenced by cropping and soil treatments. USDA Tech. Bull. 1164. U.S. Gov. Print. Office, Washington, DC.

Jenkinson, D.S., P.B.S. Hart, J.H. Rayner, and L.C. Parry. 1987. Modelling the turnover of organic matter in long-term experiments at Rothamsted. In J.H. Cooley (ed.) Soil organic matter dynamics and soil productivity. INTECOL Bull. 15. Int. Assoc. for Ecol., Athens, GA.

Jenny, A. 1941. Factors of soil formation. McGraw-Hill, New York.

Nelson, W.W., and J.M. MacGregor. 1973. Twelve years of continuous corn fertilization with ammonium nitrate or urea nitrogen. Soil Sci. Soc. Am. J. 37:583–586.

Parton, W.J., D.S. Schimel, C.V. Cole, and D.S. Ojima. 1987. Analysis of factors controlling soil organic matter levels in Great Plains grasslands. Soil Sci. Soc. Am. J. 51:1173–1179.

Paustian, K., W.J. Parton, and J. Persson. 1992. Modeling soil organic matter in organic-amended and nitrogen-fertilized long-term plots. Soil Sci. Soc. Am. J. 56:476–488.

Post, W.M., T.H. Peng, W.R. Emanuel, A.W. King, V.H. Dale, and D.L. DeAngelis. 1990. The global carbon cycle. Am. Sci. 78:310–326.

Schreiber, M.M., T.S. Abney, and J.E. Foster. 1987. Integrated pest management systems: A research approach. Res. Bull. 985. AES, Pursue Univ., W. Lafayette, IN.

Smika, D.E. 1990. Fallow management practices for wheat production in the Central Great Plains. Agron. J. 82:319–323.

Snyder, J.D., and J.A. Trofymow. 1984. A rapid accurate wet oxidation diffusion procedure for determining organic and inorganic carbon in plant and soil samples. Commun. Soil Sci. Plant Anal. 15:587–597.

Tate, K.R., D.J. Ross, and C.W. Feltham. 1988. A direct extraction method to estimate soil microbial C: Effects of experimental variables and some calibration procedures. Soil Biol. Biochem. 20:329–335.

Vance, E.D., P.C. Brookes, and D.S. Jenkinson. 1987. An extraction method for measuring soil microbial biomass C. Soil Biol. Biochem. 19:703–707.

Wilson, A.T. 1978. The explosion of pioneer agriculture: Contribution to the global CO_2 increase. Nature (London) 273:40–42.

13 Carbon and Nitrogen Mineralization as Influenced by Long-Term Soil and Crop Residue Management Systems in Australia[1]

V.V.S.R. Gupta

C.S.I.R.O.
Canberra, Australia

Peter R. Grace

CRC for Soil and Land Management
Adelaide, Australia

M.M. Roper

C.S.I.R.O.
Perth, Australia

Arable soils in Australia, especially those associated with cereal production, have suffered a severe decline in soil organic matter (SOM) (Dalal & Mayer, 1989), an important source for plant essential nutrients. In addition, loss of SOM also resulted in increased soil erosion and in turn extensive land degradation. To achieve sustainability in agriculture, it is therefore essential to improve the quantity and quality of SOM and in turn soil fertility.

Burning cereal crop residues after harvest (to reduce disease) has been a common practice. A vast amount of organic C and N is produced as crop residue. Australia produces more than 15 million t of wheat (*Triticum aestivum* L.) annually (Grace & Gessler, 1992). Using a harvest index of 0.3 and assuming cereal residues of 45% C, and with a C/N ratio of 60:1, more than 15 million t of organic C and 250 000 t of organic N are produced annually in wheat production. This is equivalent to half of the total N fertilizer usage for this industry in Australia, and is essentially a free source of slow-release fertilizer. Long-term sustainability depends on soil rather than fertilizer-

[1] Land and Water Resources Research and Development Corporation, Canberra, and Land and Water Care, C.S.I.R.O., Canberra, ACT, Australia. Current address of V.V.S.R. Gupta is: CRC for Soil and Land Management, CSIRO, Glen Osmond, Australia.

derived sources of N. It is necessary, therefore, to adopt alternative farming systems such as residue-retained, minimum, or no-till systems, and rotations with grain legumes (e.g., wheat/lupins [*Lupinus* sp.]) to improve SOM. By increasing soil organic matter levels and labile C and N stocks, the nutrient supplying potential of soil will be improved, and long-term ecological sustainability may be economically achieved.

The objective of our study was to evaluate crop residue management systems currently used in southern Australia for their ability to improve soil organic matter and C and N availability.

METHODS

Topsoil (0–10 cm) samples of red-brown earths (Alfisols) were collected prior to sowing in 1991 from adjacent on-farm treatments (Cowra and Junee) or long-term experimental plots (Rutherglen) in three key agricultural regions in southern Australia (Table 13–1). The treatments included at one or more sites include (i) residues burned (Burn), (ii) residues retained (RR), and (iii) mixed treatment (burnt/retained; MIX). For each treatment, 10 soil samples were bulked for laboratory analysis. Field moist samples were sieved (<2 mm) to remove plant and root material. Air-dried samples were used for all physical, chemical, and particulate organic matter (POM) analyses. For samples from on-farm treatments (Cowra and Junee), subsamples (~500 g) from bulk soil samples were used as replicates.

Soil pH was determined on 1:5 soil/water extracts. Soil N was determined by Kjeldahl digestion (Bremner, 1960). Soil C was determined using a LECO dry combustion analyzer. Particulate organic matter C and N was determined using the soil dispersion and wet sieving (>53 μm) method proposed by Cambardella and Elliott (1992), with mineral-associated C and N calculated by difference (i.e., Total C or N — POM C or N).

Table 13–1. Characteristics† of sites located in major dryland cereal production regions of southern Australia.

Soil property	Junee	Cowra	Rutherglen
		g kg^{-1}	
Total C	11.90	13.30	15.60
Total N	1.03	1.16	1.41
POM-C	4.02	2.80	5.10
POM-N	0.28	0.13	0.29
Microbial biomass C	0.37	0.91	0.63
Microbial biomass N	0.07	0.16	0.08
CO_2-C respired (mg kg^{-1} d^{-1})	8.46	8.63	6.55
Mineralizable C (C_0)	1.93	3.24	1.75
Mineralizable N (N_0)	0.18	0.17	0.14
Length of incubation (d)	302	84	84
pH (water)	5.30	5.60	4.90
Sand % (<250 μm)	18.20	22.80	23.80
Clay %	19.00	17.00	26.00

† Average of sampled treatments.

Microbial biomass (MB) C and N and potentially mineralizable C and N (C_0 and N_0) were determined on preincubated (72 h for microbiological stabilization) moist soil samples adjusted to 55% water-filled pore space. Microbial biomass C and N were determined using the chloroform fumigation incubation method (Jenkinson & Powlson, 1976) using a k_c of 0.40, whereas for biomass N, k_n was based on the $C_f:N_f$ ratio (Voroney & Paul, 1984). Potentially mineralizable C (CO_2-C evolved) was measured at weekly intervals using a Gowmac series 500 gas chromatograph equipped with a thermal conductivity detector, and mineralizable (NH_4^+-N and NO_3^--N in 2 M KCl extracts) was estimated after incubation at 25 °C for 80 d and calculated by linear regression (Stanford, 1969). Soil respiration was determined as CO_2—C evolved from 30-g soil samples incubated at 25 °C for 20 d (Gupta & Germida, 1988). Microbial activity was calculated using a daily respiration/biomass C ratio. Mineral N levels were measured in 2 M KCl (1:3 soil/KCl) with a Technicon autoanalyzer.

The treatment effects on MB, POM, C_0, N_0, and microbial activity levels were reported as relative values to total C and N levels for individual treatments. Treatment effects were tested using ANOVA and LSD values calculated at significance level $P = 0.05$.

RESULTS AND DISCUSSION

Average values for the chemical, physical, and microbiological properties of soils for the three sites are given in Table 13-1. The different crop residue management systems used had a significant impact on C and N levels in soil at Junee and Cowra (Table 13-2). The inclusion of grain legumes (lupins) in the cropping sequence (Rutherglen) has demonstrated clearly their beneficial role in increasing total soil C and N pools compared with the cereal-only sites. However, increased potential for the accumulation of residual NO_3^- in this system could be a drawback of the system. The accumulation of organic anions and nitrates are the main causes of net acid addition to the main pasture and cropping ecosystems of southern Australia (Helyar & Porter, 1989).

Particulate organic matter C and N levels were influenced by the overall C/N ratio of the retained material (Table 13-3). Our results show that POM-C levels at Rutherglen site increased by more than 30% where cereal and legume residues were retained in a rotation (compared with burning the

Table 13-2. Probability levels for statistical significance for treatment effects on total C and N levels in soil.

Soil property	Significance level ($P > F$)		
	Junee	Cowra	Rutherglen
Total C	<0.0001	<0.0001	NS†
Total N	<0.0001	<0.0001	<0.0005

† NS = not significant at $P \leq 0.01$.

Table 13–3. Carbon and N in particulate organic matter fraction† of soils under continuous cereal or cereal–legume rotations in dryland regions of southern Australia.

Site	Treatment‡	Crop§	POM-C Total C	POM-N Total N	POM C/N
			g kg^{-1} POM/g kg^{-1} total		
Junee	BURN	C	0.41	0.33	14.6
	RR4	C	0.29	0.25	13.9
	RR7	C	0.30	0.23	15.1
	RR15	C	0.30	0.25	12.1
Cowra	BURN	C	0.21	0.12	20.9
	MIX11	C	0.20	0.10	21.0
	RR11	C	0.22	0.11	24.8
Rutherglen	BURN	C-L	0.30	0.20	14.5
	RR8	C-L	0.40	0.25	18.3
	NT8	C-L	0.31	0.18	19.3

† Mineral-associated fraction = 1.0 − particulate organic matter.
‡ BURN = cereal residues burnt; RR4 = all residues retained for 4 (8, 11, or 15) yr; MIX = residues retained/burn (mixed management); NT = residues retained no-till.
§ C = cereal only; C-L = cereal–grain legume rotation.

cereal residues). On the other hand, in treatments where cereals were continuously grown (Junee), POM-C and N levels made up a smaller proportion of the total soil N or N in the treatments where the residues were retained than where they were burned (Table 13–3).

A stepwise increase in the size of the N_0 pool (24–88%) with time (4, 7, and 15 yr) was evident in the treatments at Junee, where cereal residues were retained (Fig. 13–1). This may have been due partly to addition of N

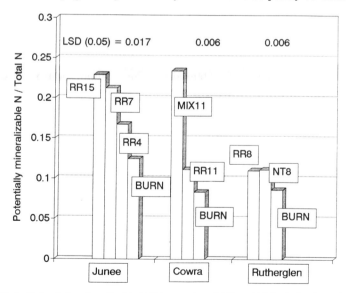

Fig. 13–1. Potentially mineralizable N (N_0, μg mineral N/total N) in dryland continuous cereal and cereal–legume rotations in southern Australia. *Burn* = cereal residues burnt; *RR4* = all residues retained for 4 yr (8, 11, or 15 yr); *MIX* = residues retained/burnt (mixed management); *NT* = residues retained no-till.

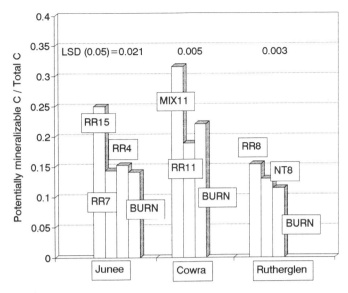

Fig. 13-2. Potentially mineralizable C (C_0, μg CO_2-C/total C) in dryland continuous cereal and cereal-legume rotations in southern Australia. *Burn* = cereal residues burnt; *RR4* = all residues retained for 4 yr (8, 11, or 15 yr); *MIX* = residues retained/burnt (mixed management); *NT* = residues retained no-till.

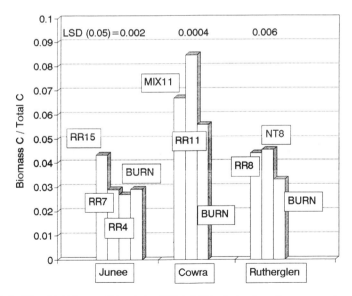

Fig. 13-3. Microbial biomass C levels (μg C/total C) in dryland continuous cereal and cereal-legume rotations in southern Australia. *Burn* = cereal residues burnt; *RR4* = all residues retained for 4 yr (8, 11, or 15 yr); *MIX* = residues retained/burnt (mixed management); *NT* = residues retained no-till.

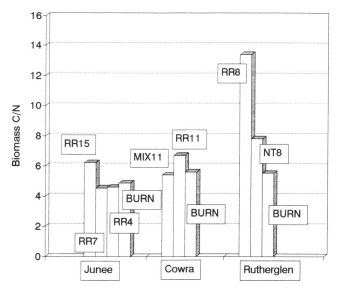

Fig. 13-4. Carbon/Nitrogen ratios of microbial biomass in dryland continuous cereal and cereal-legume rotations in southern Australia. *Burn* = cereal residues burnt; *RR4* = all residues retained for 4 yr (8, 11, or 15 yr); *MIX* = residues retained/burnt (mixed management); *NT* = residues retained no-till.

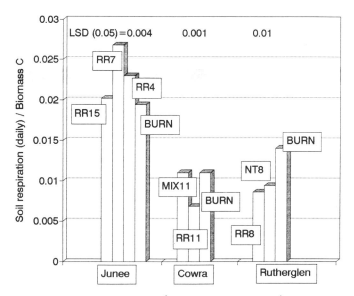

Fig. 13-5. Microbial activity (g CO_2-C kg^{-1} daily respiration/g kg^{-1} microbial biomass C) in dryland continuous cereal and cereal-legume rotations in southern Australia. *Burn* = cereal residues burnt; *RR4* = all residues retained for 4 yr (8, 11, or 15 yr); *MIX* = residues retained/burnt (mixed management); *NT* = residues retained no-till.

through fixation of N_2 by free-living bacteria using residues for energy (Roper, 1983). Increases in N_0 were not as evident in systems where residues with low C/N ratios were incorporated into the soil (Rutherglen). Long-term incorporation of residues (>7 yr) increased the size of the C_0 pool by 14 to 63% (Fig. 13-2). Prior to this time, C_0 levels remained much the same as in the treatments where residues had been removed by burning.

Increases in microbial biomass C (33-55%) only occurred in the residue-retained treatments after long-term residue incorporation (>7 yr) (Fig. 13-3). Microbial biomass C/N ratios increased under residue-retained systems (Fig. 13-4), especially in management systems where low C/N ratios residues were incorporated (Rutherglen). Where cereal residues only had been retained, microbial activity increased by 42% after 7 yr. Microbial activity was reduced by 32% when a legume was included in the rotation (Fig. 13-5).

CONCLUSIONS

Residue retention significantly increases the size (relative to total C and N) of the labile mineralizable C and N pools compared with when residue was burned. Cropping systems where cereal residues only have been retained (compared with cereal–legume rotations) are generally more responsive in terms of mineralizable C and N.

A trade-off exists between the size of the MB pool and its activity, which is dependent on the quality of the residue inputs. Continuous retention of high C/N ratio residues (cereals) increases microbial activity, but not the size of the biomass. This increased activity ensures rapid decomposition and turnover of POM and associated labile forms of residues. Legume N inputs, such as in a cereal–legume rotation, increase MB but reduce its activity (per unit MB), resulting in an accumulation of POM. Inorganic N is immobilized to maintain the relatively large biomass pool, which results in a reduction in the mineralizable labile fraction.

Many assays have been proposed as rapid indicators of soil quality in terms of OM status or fertility. Anderson and Domsch (1990) suggested the metabolic quotient for CO_2 (qCO_2; microbial activity measured as units of respiration-C per unit MB-C) as one of the indicators for quantifying changes in microbial communities. Our results indicate that a measurement of microbial activity (qCO_2) must be taken into consideration when defining the changes in the quality of OM and the health of a soil.

ACKNOWLEDGMENT

The authors thank Neil Smith and Luiz Ribeiro for technical support.

REFERENCES

Anderson, T.H., and K.H. Domsch. 1990. Application of ecophysiological quotients (qCO$_2$ & qD) on microbial biomasses from soils of different cropping histories. Soil Biol. Biochem. 22:251–255.

Bremner, J.M. 1960. Determination of N in soil by the Kjeldahl method. J. Agric. Sci. 55:11–33.

Cambardella, C.A., and E.T. Elliott. 1992. Particulate organic matter changes across a grass-land cultivation sequence. Soil Sci. Soc. Am. J. 56:777–783.

Dalal, R.C., and R.J. Mayer. 1989. Long-term trends in fertility of soils under continuous cultivation and cereal cropping in southern Queensland: II. Total organic carbon and its rate of loss from the soil profile. Aust. J. Soil Res. 24:265–279.

Grace, P.R., and P. Gessler. 1992. Strategic evaluation of rural land use in Australia using a geographic information system. p. 545–551. *In* Proc. Int. Symp. on Issues and Management of Land Resources. Feng Chia Univ., Taichung, Taiwan. 29–30 May 1992. Dep. of the Management, Fen Chia Univ., Taichung, Taiwan.

Gupta, V.V.S.R., and J.J. Germida. 1988. Distribution of microbial biomass and its activities in different soil aggregate size classes as influenced by cultivation. Soil Biol. Biochem. 20:777–789.

Helyar, K.R., and W.M. Porter. 1989. Soil acidification, its measurement and the processes involved. p. 61–100. *In* A.D. Robson (ed.) Soil acidity and plant growth. Academic Press, New York.

Jenkinson, D.S., and D.S. Powlson. 1976. The effects of biocidal treatments on metabolism in soil: V. A method for measuring microbial biomass. Soil Biol. Biochem. 8:209–213.

Roper, M.M. 1983. Field measurements of nitrogenase activity in soils amended with wheat straw. Aust. J. Agric. Res. 34:725–739.

Stanford, G. 1969. Extraction of soil organic N by autoclaving in H_2O: II. A kinetic approach to estimating the NaOH-distillable fraction. Soil Sci. 107:323–328.

Voroney, R.P., and E.A. Paul. 1984. Determination of k_C and k_n *in situ* for calibration of the chloroform fumigation–incubation method. Soil Biol. Biochem. 16:9–14.

14 Biologically Active Pools of Carbon and Nitrogen in Tallgrass Prairie Soil[1]

Charles W. Rice

Kansas State University
Manhattan, Kansas

Fernando O. Garcia

E.E.A. INTA
Balcarce, Argentina

Tallgrass prairie is one of the major ecosystems in the USA. This ecosystem is characterized by minimal external additions and losses of nutrients, such as N. Currently, much of the area formerly under tallgrass is under production agriculture. For these reasons, it is important to know the characteristics of tallgrass prairie soils to understand changes as a result of agriculture and to develop parameters for measuring soil quality. Fire and available soil water are two key determinants in maintaining the tallgrass prairie. Burning increases plant productivity by increasing the photosynthetic capacity of plants (Knapp & Seastedt, 1986). Burning also results in changes in soil temperature, soil moisture, and nutrient status. Long-term annual burning results in lower levels of soil organic matter and net N mineralization rates, but higher levels of plant productivity compared with no burning (Ojima, 1987); this indicates a change in N cycling and use. Ojima (1987) found that microbial biomass C and N were reduced by long-term annual burning (>40 yr), but were not affected by short-term burning (1–2 yr).

Decreases in soil organic N and net N mineralization rates under prolonged burning could indicate a decrease of the active pool of soil organic N. Potentially mineralizable N or N_o, as determined by the incubation method of Stanford and Smith (1972), has been suggested as an index of the capacity of the soil to supply mineralized N (Stanford, 1982; Robertson et al., 1988). If the dynamics of C and N mineralization are closely linked, then simultaneous estimation of potentially mineralizable C and N and the rates of mineralization would provide a better knowledge of soil organic mat-

[1] This research was supported by NSF grants BSR-8514327 and BSR-9011662 to Kansas State University. Contribution no. 93-179-B from the Kansas Agric. EWxp. Stn., Manhattan, KS.

ter turnover (Paul & Juma, 1981; van Veen et al., 1984; Robertson et al., 1988; Nadelhoffer, 1990).

Our objective was to determine the size and the ratio of biologically active pools of C and N in tallgrass prairie soil under different burning and N fertilization management. Under these conditions, we would expect an optimum soil quality. The evaluation of the size and ratios of biologically active pools of C and N could contribute to a definition of soil quality (see Chapter 8, this book).

MATERIALS AND METHODS

The experimental site is located on the Konza Prairie Research Natural Area, Kansas. Konza Prairie is a remnant of unplowed tallgrass prairie. The vegetation is mainly dominated by C_4 grasses such as big bluestem (*Andropogon gerardii* Vit.), little bluestem (*A. scoparius* Michx.), and indiangrass [*Sorghastrum nutans* (L.) Nash]. The experiment was started in 1986 on an area mapped as Irwin silty clay loam (fine, mixed, mesic Pachic Argiustolls). The experimental design was a split-split plot—the main plot in a randomized complete block design with four blocks, the main plot corresponds to burning (annually burned and unburned), the subplot to mowing (annually mowed and unmowed), and the sub-subplot to N fertilization (control and N fertilized). Burning occurred in late spring, usually the last week of April. The plots normally were mowed and raked once a year in early June. Nitrogen was applied 7 to 10 d after burning, usually the first week of May, at a rate of 10 g N m^{-2} as NH_4NO_3. In this study, we considered the effects of annual burning and N fertilization in the unmowed prairie.

Soil samples (0 to 5-cm depth) were obtained from each replicate on 29 Mar. 1991 to determine total organic C (TOC) and N (TON), and microbial biomass C (MBC) and N (MBN). Total organic C was determined by dry combustion using a WR-12 Leco Carbon Analyzer (LECO Corp., St. Joseph, MI). Total organic N was determined by salicylic–sulfuric acid digestion (Bremner & Mulvaney, 1982) and measuring the N content of the digests on an Alpkem Autoanalyzer (Alpkem Corp., Bull. A303-S021). Microbial biomass C and N were determined by the fumigation–incubation method (Jenkinson & Powlson, 1976) and calculated as suggested by Voroney and Paul (1984).

Pools of potentially mineralizable C and N (PMC and PMN) and their rates of mineralization were determined by a modification of the method proposed by Cabrera and Kissel (1988a) (Garcia, 1992). We used undisturbed soil samples in which mineralized C and N were measured at controlled soil water potentials. Three undisturbed soil samples were obained from each replicate of the burning and N-fertilized treatment combinations on 6 Apr. 1991 before burning. Cores of polyvinyl chloride (PVC) (5.08-cm diam., 10-cm height) were driven 5 cm into the soil and then carefully removed. The samples were incubated in 940-mL mason jars at 35 °C over a 218-d period. Mineralized N was measured at 7, 14, 28, 42, 62, 84, 112, 147, 175, and

218 d of incubation by leaching with 0.01 M $CaCl_2$ and determining inorganic N in the leachate. Between leaching events, accumulated CO_2-C was measured by gas chromatography. Soil cores were disassembled at the end of the last leaching (218 d) and MBC, MBN, and TOC were measured as indicated above.

Potentially mineralizable pools of C and N and mineralization rates were estimated by kinetic analysis (Stanford & Smith, 1972; Brunner & Focht, 1984). Carbon mineralization data were described by first-order kinetics by fitting for each core a one-pool model of the form

$$C_m = PMC\,[1 - \exp(-kt)]$$

where C_m (g CO_2-C g^{-1} soil) is mineralized C at time t, PMC (g CO_2-C g^{-1} soil) is potentially mineralizable C, k (d^{-1}) is the rate constant of mineralization, and t (d) is time.

Nitrogen mineralization data were fitted to a mixed-order model (Brunner & Focht, 1984; Bonde & Lindberg, 1988). The general form of the mixed-order model is

$$N_m = PMN[1 - \exp(-k_1 t - k_2 t^2/2)]$$

where N_m (mg N kg^{-1} soil) is mineralized N at time t, PMN (mg N kg^{-1} soil) is potentially mineralizable N, k_1 (d^{-1}) are the rate constants of mineralization, and t (d) is time.

We fractionated the total soil organic C and N into discrete pools of different biological activity using the values of MBC, PMC, and TOC; and MBN, PMN, and TON. Microbial biomass pools are considered the most labile fractions of soil C and N. The fractions of PMC and PMN not accounted for by the decreases in MBC and MBN were defined as nonmicrobial biomass active pools. Nonmicrobial biomass active pools of C and N (NBA-C and NBA-N) were calculated by subtracting the changes in MBC and MBN through incubation from PMC and PMN, respectively. The microbial biomass and the nonmicrobial biomass active pools were subtracted from the total soil organic C and N to estimate the stable soil organic C and N pools (SOC and SON). The stable pools would have a slower rate of turnover than active pools.

RESULTS

The size of the pools of C and N in tallgrass prairie soil were not significantly affected after 6 yr of annual burning and N fertilization (Table 14–1). Burning tended to decrease TOC, but this trend was hardly observed in TON. Nitrogen fertilization tended to increase TOC in the unburned prairie, and TON in the unburned or burned prairie.

Microbial biomass C and N tended to be higher in the burned than in the unburned prairie. Addition of N resulted in similar levels fo MBC and

Table 14-1. Levels of C and N in the soil organic matter, microbial biomass, and mineralizable fractions.

Treatment	TOC†	PMC	MBC	TON	PMN	MBN
	— g C kg^{-1} —			— g N kg^{-1} —		
Unburned	31.0	8.86	0.98	2.69	0.35	0.29
Unburned + N	33.1	7.45	1.10	2.81	0.41	0.30
Burned	30.9	10.33	1.20	2.62	0.42	0.32
Burned + N	30.2	9.66	1.10	2.71	0.42	0.29
	Analysis of variance					
	$Pr > F$					
Burning	0.378	0.347	0.652	0.618	0.604	0.739
N fertilization	0.551	0.338	0.922	0.155	0.426	0.778
Burning × N fertilization	0.244	0.730	0.498	0.785	0.432	0.479

† TOC = total organic carbon; PMC = potentially mineralizable carbon; MBC = microbial biomass carbon; TON = total organic nitrogen; PMN = potentially mineralizable nitrogen; MBN = microbial biomass nitrogen.

MBN for burned or unburned prairie. These results correspond to the beginning of the growing season when microbial biomass values are highest in tallgrass prairie (Garcia, 1992). Our research indicates that MBC and MBN are highly dynamic in tallgrass prairie and that the temporal dynamics are significantly affected by burning and N additions.

Burning also tended to increase PMC and PMN pools. No significant differences were detected among the rate constants for C mineralization. The first rate constant of N mineralization, k_1, was significantly higher with N fertilization ($k_1 = 0.0076$) than the unfertilized control ($k_1 = 0.0046$) when averaged across burned treatments. This indicates faster turnover of PMN in the fertilized treatments. The second rate constant, k_2, was significantly higher in the unburned treatment ($k_2 = 0.000\ 131$) than in the burned treatment ($k_2 = 0.000\ 063$). This could indicate greater initial immobilization of N in the unburned prairie soil during incubation.

Nonbiomass active pools of C tended to be greater in burned than unburned prairie and represent a greater proportion of TOC. Nitrogen additions tended to decrease the size of NBA-C and NBA-N, especially in unburned prairie.

A trend for greater pools of SOC and SON in unburned than burned prairie was apparent. With a tendency for greater active pools in burned prairie and no differences in total soil organic pools, we could expect a trend for more stable pools in unburned prairie. A greater proportion of soil organic matter in stable pools would result in a greater conservation of nutrients. Addition of N tended to increase the size of the stable C pool in the unburned prairie.

DISCUSSION

Using the Century model, Ojima et al. (1990) predicted significant decreases in soil organic matter levels after 10 to 12 yr of annual burning.

However, 6 yr of annual burning on tallgrass prairie have had no significant effect on the size of TOC and TON, but have tended to increase the size of active pools (PMC, PMN, MBM-C, NBA-C, and NBA-N) at the expense of the stable pools. Simulation modeling by Risser and Parton (1982) indicated that a buildup of N of 0.34 g N m^{-2} yr^{-1} in unburned, and a net loss of 1.09 g N m^{-2} yr^{-1} in burned prairie could occur by the 6th yr of burning, but they did not find significant differences in TON after 6 yr. The lack of significant differences between burning treatments in our study is probably because the experiment was evaluated at a transition period between the short- and long-term responses to burning. Short-term responses to burning include greater active N pools and N mineralization rates. On the other hand, long-term burning results in a decrease of soil organic matter and N mineralization rates (Ojima, 1987; Ojima et al., 1990).

Microbial biomass C and N of tallgrass prairie are higher than those of agricultural ecosystems that range from 0.2 to 0.7 g C kg^{-1}, and 0.04 to 0.16 g N kg^{-1} for MBC and MBN, respectively (Carter & Rennie, 1982; Schimel et al., 1985; McGill et al., 1986; Schimel, 1986; Insam, 1990; Patra et al., 1990). Compared with other grasslands, the tallgrass prairie has lower or similar estimates of MBC, but values of MBN are generally higher, resulting in a narrower C/N ratio of the microbial biomass in tallgrass prairie than other grasslands (Schimel et al., 1985; Schimel, 1986; Woods, 1989; Tate et al., 1991). The proportion of TOC as MBC in tallgrass prairie (ranges from 0.4 to 6%) is similar to that in agroecosystems, but the proportion of TON as MBN is higher than that in agroecosystems (ranges from 2 to 7%) (Fig. 14–1) (Lynch & Panting, 1980; Jenkinson & Ladd, 1981; Robertson et al., 1988; Insam, 1990). It is important to consider the variation in microbial biomass when assessing soil quality. We have measured a 50% change in seasonal values of MBC and MBN, which would significantly change the ratios of microbial biomass to other soil parameters.

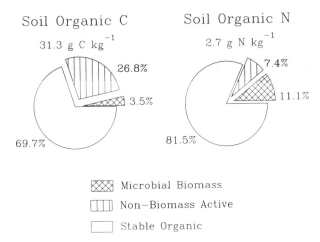

Fig. 14–1. Pools of organic C and N in tallgrass prairie.

Table 14-2. Relationships between potentially mineralizable C (PMC), N (PMN), total organic C (TOC), and N (TON), and microbial biomass C (MBC) and N (MBN) from tallgrass prairie soil.

Treatment	PMC/PMN	PMC/TOC	MBC/PMC	PMN/TON	MBN/PMN
Unburned	25.3	0.29	0.11	0.14	0.77
Unburned + N	18.2	0.23	0.13	0.15	0.63
Burned	24.6	0.33	0.12	0.17	0.69
Burned + N	22.8	0.32	0.12	0.17	0.67

The PMC and PMN pools accounted for approximately 29 and 15%, respectively, of the total organic C and N pools, respectively (Table 14-2). Both proportions are slightly greater than values previously reported for agricultural ecosystems (Bonde et al., 1988; Robertson et al., 1988). Studies in agroecosystems showed a range from 5 to 18% of total N as PMN (Campbell & Souster, 1982; Bonde et al., 1988; Cabrera & Kissel, 1988b; Robertson et al., 1988). For the conditions of our study, we found that PMN was correlated to TOC and TON contents (Garcia, 1992).

Microbial biomass C and N averaged 12 and 75% of PMC and PMN, respectively. The decrease in MBC and MBN during the incubation contributed 9 and 80% to PMC and PMN, respectively. In agricultural ecosystems, the proportions of MBC and MBN in PMC and PMN vary from 2 to 36% and 18 to 89%, respectively (Bonde et al., 1988; Boyle & Paul, 1989). These values suggest that MBN contributes significantly to PMN in most ecosystems.

The greater proportion of active pools of C and N for TOC and TON suggests faster soil organic matter turnover in the burned than unburned prairie. This is supported by lower PMC/PMN ratios, higher first-rate constant of mineralization (k_1), and lower second-rate constant (k_2) for the burned prairie. Faster turnover may translate to greater N availability (Garcia, 1992). Values of PMC/PMN ratio were in agreement with those reported for rangeland (Schimel et al., 1985). Robertson et al. (1988) reported lower PMC/PMN, 11.2, for a long-term field experiment that had been annually fertilized and cropped to cereals. The size of the mineralizable pool of this arable soil is smaller than that of tallgrass prairie, but its narrow C/N ratio indicates a greater turnover potential. Doran and Smith (1987) suggested that changes in nutrient cycling, and thus nutrient availability, are associated with changes in the relative quantity and activity of labile and stable organic matter pools. Undisturbed ecosystems, such as tallgrass prairie or no-till agriculture, could result in a greater conservation of N by immobilization into less active organic pools.

In tallgrass prairie, the soil receives inputs from primary production, and soil organic matter interacts with primary production by providing needed nutrients. This relationship is regulated by environmental factors, such as moisture and temperature, and disturbances, such as fire or grazing. Under such a natural ecosystem, we would expect to find a situaiton of high soil quality. Values of MBC/TOC, MBN/TON, PMC/TOC, and PMN/TON calculated from our research represent an equilibrium that results in sustained primary productivity and conservation of nutrients. In agricultural soils, we would expect smaller C and N pools, but greater proportions of active soil C and N pools, and/or higher turnover rates (Beauchamp et al., 1986). Management systems, such as no-till, that maintain the conservation of residues would increase the relative size of active pools with a slower turnover (Rice et al., 1986). Using these ratios may be useful for assessing changes in soil quality.

ACKNOWLEDGMENTS

We are grateful to R. Ramundo and the personnel of Konza Prairie Research Natural Area for help in the field experiment. This research was supported by NSF grants BSR-8514327 and BSR-9011662 to Kansas State University. Contribution no. 93-179-B from the Kansas Agric. Exp. Stn., Manhattan, KS.

REFERENCES

Beauchamp, E.G., W.D. Reynolds, D. Brasche-Villeneuve, and K. Kirby. 1986. Nitrogen mineralization kinetics with different soil pretreatments and cropping histories. Soil Sci. Soc. Am. J. 50:1478–1483.

Bonde, T.A., and T. Lindberg. 1988. Nitrogen mineralization kinetics in soil during long-term aerobic laboratory incubations: A case study. J. Enviorn. Qual. 17:414–417.

Bonde, T.A., J. Schnurer, and T. Rosswall. 1988. Microbial biomass as a fraction of potentially mineralizable nitrogen in soils from long-term field experiments. Soil Biol. Biochem. 20:447–452.

Boyle, M., and E.A. Paul. 1989. Carbon and nitrogen mineralization kinetics in soil previously amended with sewage sludge. Soil Sci. Soc. Am. J. 53:99–103.

Bremner, J.M., and C.S. Mulvaney. 1982. Nitrogen—total. p. 595–624. In A.L. Page et al. (ed.) Methods of soil analysis. Part 2. 2nd ed. Agron. Monogr. 9. ASA and SSSA, Madison, WI.

Brunner, W., and D.D. Focht. 1984. Deterministic three-half-order model for microbial degradation of added carbon substrates in soil. Appl. Environ. Microbiol. 47:167–172.

Cabrera, M.L., and D.E. Kissel. 1988a. Potentially mineralizable nitrogen in disturbed and undisturbed soil samples. Soil Sci. Soc. Am. J. 52:1010–1015.

Cabrera, M.L., and D.E. Kissel. 1988b. Evaluation of a method to predict nitrogen mineralized from soil organic matter under field conditions. Soil Sci. Soc. Am. J. 52:1027–1031.

Campbell, C.A., and W. Souster. 1982. Loss of organic matter and potentially mineralizable nitrogen from Saskatchewan soils due to cropping. Can. J. Soil Sci. 62:651–656.

Carter, M.R., and D.A. Rennie. 1982. Changes in soil quality under zero tillage farming systems: Distributions of microbial biomass and mineralizable C and N potentials. Can. J. Soil Sci. 62:587–597.

Doran, J.W., and M.S. Smith. 1987. Organic matter management and utilization of soil and fertilizer nutrients. p. 53–72. In R.F. Follet et al. (ed.) Soil fertility and organic matter as critical components of production systems. Spec. Publ. 19. ASA and SSSA, Madison, WI.

Garcia, F.O. 1992. Carbon and nitrogen dynamics and microbial ecology in tallgrass prairie. Ph.D. diss. Kansas State Univ., Manhattan (Diss. Abstr. 92-35625).

Insam, H. 1990. Are the soil microbial biomass and basal respiration governed by the climatic regime? Soil Biol. Biochem. 22:525–532.

Jenkinson, D.S., and J.N. Ladd. 1981. Microbial biomass in soil: Measurement and turnover. p. 415–471. In E.A. Paul and J.N. Ladd (ed.) Soil biochemistry. Vol. 5. Marcel Dekker, New York.

Jenkinson, D.S., and D.S. Powlson. 1976. The effects of biocidal treatments on metabolism in soil: V. A method for measuring soil biomass. Soil Biol. Biochem. 8:209–213.

Knapp, A.K., and T.R. Seastedt. 1986. Detritus accumulation limits productivity of tallgrass prairie. BioScience 36:662-668.

Lynch, J.M., and L.M. Panting. 1980. Cultivation and the soil biomass. Soil Biol. Biochem. 12:29-33.

McGill, W.G., K.R. Cannon, J.A. Robertson, and F.D. Cook. 1986. Dynamics of soil microbial biomass and water-soluble organic C in Breton L after 50 years of cropping to two rotations. Can. J. Soil Sci. 66:1-19.

Nadelhoffer, K.J. 1990. Microlysimeter for measuring nitrogen mineralization and microbial respiration in aerobic soil incubations. Soil Sci. Soc. Am. J. 54:411-415.

Ojima, D.S. 1987. The short-term and long-term effects of burning on tallgrass prairie ecosystem properties and dynamics. Ph.D. diss. Colorado State Univ., Ft. Collins (Diss. Abstr. 87-25646).

Ojima, D.S., W.J. Parton, D.S. Schimel, and C.E. Owensby. 1990. Simulated impacts of annual burning on prairie ecosystems. p. 118-132. In S.L. Collins and L.L. Wallace (ed.) Fire in North American tallgrass prairies. Univ. of Oklahoma Press, Norman, OK.

Patra, D.D., P.C. Brookes, K. Coleman, and D.S. Jenkinson. 1990. Seasonal changes in soil microbial biomass in an arable soil and a grassland soil which have been under uniform management for many years. Soil Biol. Biochem. 22:739-742.

Paul, E.A., and N.G. Juma. 1981. Mineralization and immobilization of soil nitrogen by microorganisms. In F.E. Clark and T. Rosswall (ed.) Terrestrial nitrogen cycles. Ecol. Bull. 33:179-195.

Rice, C.W., M.S. Smith, and R.L. Blevins. 1986. Soil nitrogen availability after long-term continuous no-tillage and conventional tillage corn production. Soil Sci. Soc. Am. J. 50:1206-1210.

Risser, P.G., and W.J. Parton. 1982. Ecosystem analysis of the tallgrass prairie: Nitrogen cycle. Ecology 63:1342-1351.

Robertson, K., J. Schnurer, M. Clarholm, T. Bonde, and T. Rosswall. 1988. Microbial biomass in relation to C and N mineralization during laboratory incubations. Soil Biol. Biochem. 20:281-286.

Schimel, D.S. 1986. Carbon and nitrogen turnover in adjacent grassland and cropland ecosystems. Biogeochemistry 2:345-357.

Schimel, D.S., D.C. Coleman, and K.A. Horton. 1985. Soil organic matter dynamics in prairie rangeland and cropland toposequences in North Dakota. Geoderma 36:201-214.

Stanford, G. 1982. Assessment of soil nitrogen availability. p. 651-688. In F.J. Stevenson (ed.) Nitrogen in agricultural soils. Agron. Monogr. 22. ASA, CSSA, and SSSA, Madison, WI.

Stanford, G., and S.J. Smith. 1972. Nitrogen mineralization potentials of soils. Soil Sci. Soc. Am. Proc. 36:465-472.

Tate, K.R., D.J. Ross, A. Ramsay, and K. Whale. 1991. Microbial biomass and bacteria in two pasture soils: An assessment of measurement procedures, temporal variations, and the influence of P fertility status. Plant Soil 132:233-241.

van Veen, J.A., J.N. Ladd, and M.J. Frissel. 1984. Modelling C and N turnover through the microbial biomass in soil. Plant Soil 76:257-274.

Voroney, R.P., and E.A. Paul. 1984. Determination of k_C and k_N in situ for calibration of the chloroform fumigation-incubation method. Soil Biol. Biochem. 16:-9-14.

Woods, L.E. 1989. Active organic matter distribution in the surface 15 cm of undisturbed and cultivated soil. Biol. Fertil. Soils 8:271-278.

15 A Method to Determine Long-Term Anaerobic Carbon and Nutrient Mineralization in Soils

Karen Updegraff, Scott D. Brigham, John Pastor, and Carol A. Johnston

Natural Resources Research Institute
University of Minnesota
Duluth, Minnesota

Long-term nutrient mineralization potentials, particularly of N, have become a commonly used measure of soil productivity. Aerobic laboratory incubations have been used to estimate N availability in soils (Binkley & Hart, 1989) and to determine the impact of management practices and environmental parameters on N availability (Stanford & Smith, 1972; Cassman & Munns, 1980; Bonde & Rosswall, 1987; Bonde & Lindberg, 1988). Aerobic incubations carried out in microlysimeter containers allow the concomitant measurement of both N and C mineralization potentials (Nadelhoffer, 1990), which are closely linked (McGill & Cole, 1981). However, different and complex suites of factors control C and N mineralization under saturated (Mah et al., 1977; Reeburgh, 1983; Moore & Knowles, 1989) and unsaturated conditions (Blet-Charaudeau et al., 1990; Schimel et al., 1990). Short-term anaerobic incubations have been used to investigate either N (e.g., Waring & Bremner, 1964; Powers, 1980) or C mineralization (e.g., Yavitt et al., 1987), or both simultaneously (Gale & Gilmour, 1988), but long-term anaerobic N and C kinetics have not, to our knowledge, been simultaneously described. Parallel incubation methods are needed to produce comparable aerobic and anaerobic mineralization data.

We developed an anaerobic incubation technique that closely parallels the long-term aerobic incubation procedure of Nadelhoffer (1990). Periodic leachings and gas flux measurements permit estimation of nutrient and C mineralization rates. We first describe this anaerobic method and then present results from parallel aerobic and anaerobic incubations of three wetland soils. The method is sufficiently robust to be used as a comparative index of soil quality for a broad range of organic and mineral wetland soils.

METHODS

Incubation Setup

Soil samples at field moisture are hand-mixed, removing large stones and roots, under a N_2 atmosphere in a glovebox. A portion of the sample is removed for the anaerobic incubation (see below); the remainder is used for aerobic incubation and soil characterization (soil moisture, total N and C, etc.).

Both aerobic (Nadelhoffer, 1990) and anaerobic soils are incubated in 150-ml polypropylene Falcon filter units (model 7102, Becton, Dickinson & Co., Cockeysville, MD). However, the factory-installed nitrocellulose filters are N-contaminated and must be replaced with nonbiodegradable glass fiber filters (e.g., Fisherbrand G2 filters). Due to the long incubation periods, extreme care is necessary to minimize contamination of the samples with nutrients or C. All components should be soaked in 10% HCl and rinsed with deionized water prior to use. Untreated glasswool released large amounts of CO_2 during anaerobic incubation, probably due to carbonate contamination; this contamination is eliminated by acid-washing. The use of ultrapurified water also removes a possible source of trace organic contaminants.

Aerobic Incubation Method. Aerobic incubations conformed to the method of Nadelhoffer (1990), with minor deviations as described below. The aerobic incubations are assembled after the anaerobic incubations, to avoid exposure to air of the anaerobic samples. Organic soils have low bulk densities compared with mineral upland soils, therefore, we used 2 to 20-g dry weight equivalent of soil, rather than 50 g. In addition, the headspace sampling procedure varies slightly in that the tubes connecting the top and bottom chambers are removed (leaving lid apertures open) except during headspace incubations. To obtain gas flux estimates, the top and bottom chambers are connected and sealed using Tygon tubing. The chambers are then purged with CO_2-free air (scrubbed with 2 M KOH) for sufficient time to eliminate measurable CO_2 (at least 2 min) and incubated overnight. A 10-mL syringe is used to mix the air within the chambers before sampling for direct CO_2 analysis (see below).

Anaerobic Incubation Method. The complete anaerobic incubation assembly is shown in Fig. 15-1. The sample cups of the filter units are separated from the bottom (receiver) units and lined with a double layer of nylon (l) (cut from stockings), held in place by a split retaining ring (H) (cut from polypropylene containers) and a layer of glass wool (G). The incubation materials, including samples, sand, Mason jars and lids, electronic scale, beaker and spatula, distilled water, waste receptacles, and sample cups are assembled in the anaerobic glovebox. Sufficient sample to half-fill the sample cut is weighed into a beaker, then 20- to 50-g acid-washed quartz sand is added and mixed thoroughly. For organic or very fine-textured soils, a higher sand/soil ratio is needed to facilitate leaching. The mixed soil samples are transferred to the cups (F) and covered with another layer of nylon (J). The

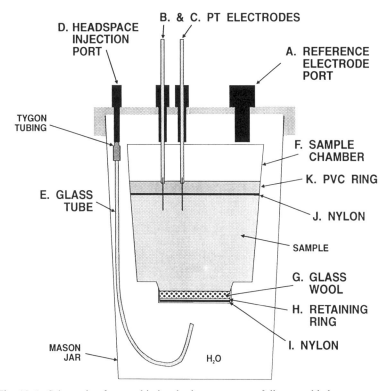

Fig. 15-1. Schematic of anaerobic incubation apparatus, fully assembled.

nylon fabric is held in place with a rigid PVC retaining ring (K) (machined from 5-cm PVC pipe couplings). The nylon and glass wool prevent the samples from floating out of the cups when completely submerged in water. The samples are sealed into N_2-flushed 500 mL (pint) Mason jars prior to the initial leaching (see Nutrient Mineralization, below); the bottom halves of the filter units are stored separately. The anaerobic samples must be saturated after the week 0 leaching by partially filling the jars with degassed distilled water, then slowly lowering the samples into the water to minimixe the trapping of bubbles. The jars are then completely filled with water and sealed.

The Mason jar dome lids tend to corrode, and were replaced with machined stainless steel lids and silicone rubber gaskets. Two 5-mm holes (Fig. 15-1, B and C) allow the installation of Pt electrodes for the measurement of Eh (Faulkner et al., 1989). The stopper in hole D connects to a glass j-shaped tube (E) for headspace injection (see below). Aperture A is a 12-mm port for the introduction of a calomel reference or pH electrode. Silicone or neoprene septum stoppers are used to seal the ports and for gas sampling (from port A).

Gas Flux Measurement

We define C mineralization as the conversion of organic to inorganic forms of C, i.e., CH_4 or CO_2. Incompletely mineralized dissolved C fractions, such as carbohydrates or fatty acids, may be measured in the water if desired. For our purposes, C mineralization is estimated by periodic measurement of gas flux: CO_2 only for aerobic samples (Nadelhoffer, 1990), CO_2 plus CH_4 for anaerobic samples.

To measure anaerobic gas flux, a headspace is created by injecting 100 mL He through septum D. The j-shaped tube forces the gas to bubble upward, mixing the sample water and dissolved gases. Displaced water is expressed through a two-way needle inserted through septum A, and transferred via tygon tubing to appropriate containers. The volume of expressed water represents actual headspace volume, including trapped bubbles. Sequential sampling has shown that at least 4 h are required for dissolved gases in the water to equilibrate with the headspace. After 4 h, samples of the headspace are drawn through septum A with a syringe and the puncture resealed with silicone grease. Immediate gas chromatographic analysis of headspace gases is preferable; however, 2.5-mL headspace samples may be stored in evacuated 2-mL blood collection tubes. The overpressurization of the tubes minimizes contamination with air. Gas standards stored in these tubes for ca. 3 mo showed no significant change in concentration. As purchased, the pre-evacuated tubes are contaminated with CH_4 (> 10 nL L^{-1}), which can be eliminated by removing and boiling the stoppers for at least 1 h in deionized water, then reevacuating the tubes.

Carbon dioxide is determined by thermal conductivity on 2 m by 3 mm Porapak-Q column at 65 °C. After each sequence of three injections, the oven temperature is briefly raised to 120 °C to drive off water. Methane is analyzed at 50 °C on a 2 m by 3 mm Porapak-Q column using a flame ionization detector. Certified gas standards (Scott Specialty Gases, Plumsteadville, PA) are used to construct standard curves for both CO_2 and CH_4. Because of the wide variation in CH_4 emissions from different soils, multiple-range standard curves may be required for proper resolution of the full range of samples.

Carbon mineralization is calculated in one of two ways:

1. By extrapolating a 24-h flux rate to the entire applicable incubation period. Sequential sampling showed that after an initial flush, both CO_2 and CH_4 emissions were roughly constant up to 6 wk after leaching.
2. By taking a single headspace sample at the end of an incubation period, whose concentration represents the total flux during that period. This is comparable to the method used to estimate N mineralization rates (see below).

Method 1 is used for aerobic samples, which are left open to ambient air and only sealed during headspace sampling (see Aerobic Incubation Method, above). However, high dissolved gas concentrations in anaerobic samples, sample disturbance, and analytical error may distort long-term flux estimates based on 24-h point measurements. Method 2 minimizes the impact of disturbance and measurement error, and offers a more integrated long-term measure of C flux. Therefore, Method 2 is preferable for the measurement of gas flux in anaerobic samples.

Nutrient Mineralization

The basic leaching procedure is identical for aerobic and anaerobic incubations, except that anaerobic leachings are performed in a glovebox under a N_2 atmosphere. The assembled filter units, with the lids removed, are connected to a valve manifold. One hundred mL of 0.01 M $CaCl_2$ are then added to the samples in three aliquots, allowing 5 min of eqilibration time after each addition before applying a vacuum. This may be followed by the addition of 25 mL of a Hoagland's nutrient solution (minus the nutrients of interest) to alleviate nutrient limitation. Nutrient additions have been standard in aerobic incubations (Stanford & Smith, 1972; Nadelhoffer, 1990), but omitting nutrient additions may provide data more comparable to field mineralization, while allowing measurement of several nutrients simultaneously (e.g., N and P).

After the final addition of extractant (or nutrient solution), the samples are drained by applying a vacuum of approximately 0.08 MPa for at least 10 s or until the visible drainage rate approaches zero. The time required varies widely depending on soil texture. However, if consistent drainage criteria are applied, the final moisture content for most soils varies by less than 20%. The leachates are collected in the bottom chamber and then expelled under pressure via the siphon port, with rinsing, into flasks. Leachates are brought to 200 mL with H_2O, subsampled, and frozen until analysis. The results of the initial (week 0) leaching are not included in calculations of long-term mineralization potential. Subsequent leachings take place at 2, 4, 8, 12, 16, 22, and 30 wk, and every 10 wk thereafter.

Additional steps are necessary to leach the anaerobic samples. The sample cups must be removed from their jars and installed on the corresponding receiver units. Water retained in the samples is removed by suction and returned to the sample jars under N_2 pressure. All of the incubated water is then transferred to labeled empty containers, and the sample jars are partially refilled with fresh degassed water. After the removal of excess water, the samples are leached as above. After the final drainage, the sample cups are lifted off of the receiver units and returned to the partly filled jars. The water is allowed to percolate upward until the soil is completely saturated, to minimize the trapping of bubbles. Then the jar is completely filled with water and sealed.

Recovered (incubated) water volume is measured at each leaching. In addition, the volume of water displaced as headspace for gas measurements must be accounted for when calculating mineralized quantities of C and nutrients. If leachings take place within 48 h after headspace creation, one can reasonably assume that the nutrient concentration measured in the recovered water represents that of the water removed as headspace. The incubated water may contain sediment from the samples. To clean the water for analysis and quantify sediment loss, water subsamples are filtered through glass-fiber syringe filters (Gelman type A/E), and the filters are dried and weighted. Therefore, sediment-loaded waters must be analyzed separately, otherwise they may be composited with the leachates. We analyze for dissolved N using Cd reduction for NO_3-N and a salicilate–nitroprusside method for NH_4-N (Quickchem methods no. 12-107-04-01-A and 10-107-06-2-A, 1986, Lachat Instruments, Mequon, WI).

RESULTS AND DISCUSSION

We present a brief summary of results from 80-wk incubations at 30 °C of wetland soils from Voyageurs National Park in northern Minnesota. Ten replicate samples each were obtained from the Oa horizon (0–15 cm) of a sedge meadow in a drained beaver pond (Typic Haplaquept) and 0 to 15 and 80 to 100 cm in a spruce (*Picea* sp.)-*Sphagnum* bog (Typic Borohemist). Some properties of these samples are listed in Table 15-1. Total N and C were measured by combustion on a Leco CHN analyzer (Leco Corp., St. Joseph, MI). Proximate C analyses were performed using sequential extractions with dichloromethane, hot water, and sulfuric acid (McClaugherty et al., 1985).

Hoagland's solution minus N (Stanford & Smith, 1972) was added to samples during leaching, in conformity with the traditional method. However, the long-term impact of nutrient additions appears minimal in these soils: subsequent long-term incubations of the same soils without nutrient additions had mineralized similar quantities of C and N after 22 wk.

Modeling Mineralization Data

Various decomposition models are available to fit sequential mineralization data. Aerobic N mineralization was described as a first-order exponential decay function by Stanford and Smith (1972), and may be generalized as:

Table 15-1. Properties of samples.†

	Sphagnum surface	*Sphagnum*, 1 m	Sedge meadow
%C	45.5 (1.3)	50.5 (1.0)	18.4 (5.7)
%N	1.3 (0.2)	1.7 (0.1)	1.2 (0.3)
pH	3.7 (0.1)	3.8 (0.1)	5.8 (0.1)
C/N ratio	35.4 (6.2)	30.4 (2.0)	14.8 (1.9)
%NPE	9.6 (0.9)	13.4 (1.0)	12.1 (1.8)
%WS	9.1 (0.9)	4.2 (1.4)	7.6 (0.4)
%AS	50.0 (5.7)	35.8 (3.7)	22.5 (4.0)
%Lignin	29.9 (5.7)	45.6 (3.3)	46.8 (3.7)
% Tannin	0.45(0.15)	0.30(0.14)	0.06(0.34)

† NPE = nonpolar extractives (waxes, resins); WS = water-soluble extractives (carbohydrates, nonstructural proteins); AS = acid-soluble extractives (cellulose, hemicellulose).
‡ SD in parentheses.

$$X_1 = X_0 \, e^{-kt} \qquad [1]$$

where X_1 is cumulative C or N release to time t, X_0 is the total pool or readily mineralizable C or N, and k is the instantaneous release rate. Where the data did not conform to the first-order decay model, we were generally able to fit a modified first-order model (Bonde & Roswall, 1987):

$$X_t = (X_1 e^{-kt}) + \beta t \qquad [2]$$

where X_1 is the labile pool of C or N with release rate k, and β is the release rate of a recalcitrant fraction. This equation describes the initial exponential decay of a labile fraction, and the subsequent linear decay of a recalcitrant fraction. Note that X_1 is not equivalent to X_0. Where mineralization rates remained constant throughout the incubation (anaerobic C from the *Sphagnum* peats), only a linear function could be fit to the data. This was of the form:

$$X_t = \delta t + b \qquad [3]$$

where δ is the release rate constant; and b is the intercept. Table 15–2 summarizes the various models used to describe N and C mineralization for these incubations. It demonstrates that a change in a single controlling factor (soil aeration) can qualitatively change C and N mineralization kinetics in soils.

Sample Redox Status

We monitored sample Eh at each leaching and gas sampling using pairs of copper-platinum electrodes (Faulkner et al., 1989) inserted through ports B and C (Fig. 15–1). Coefficients of variation for duplicate electrodes often exceeded 100%, particularly in the *Sphagnum* peats. During 80 wk of incubation, mean Eh in the sedge soils rose gradually from a low of -182 mV to $+12$ mV, whereas in the *Sphagnum* peats mean Eh ranged from $+29$ to $+220$ mV, with a slight declining trend.

Nitrogen Mineralization

The parallel methods allowed us to assess the impact of aeration status on soil N turnover rates. Nitrogen mineralization was up to three times higher

Table 15–2. Models used to fit N and C mineralization data. The term *Exponential* refers to Eq. [1], *Modified* to Eq. [2], and *Linear* to Eq. [3] in the text.

	Sphagnum surface	*Sphagnum*, 1 m	Sedge meadow
N			
Aerobic	Exponential	Exponential	Exponential
Anaerobic	Modified	Modified	Exponential
C			
Aerobic	Modified	Modified	Modified
Anaerobic	Linear	Linear	Modified

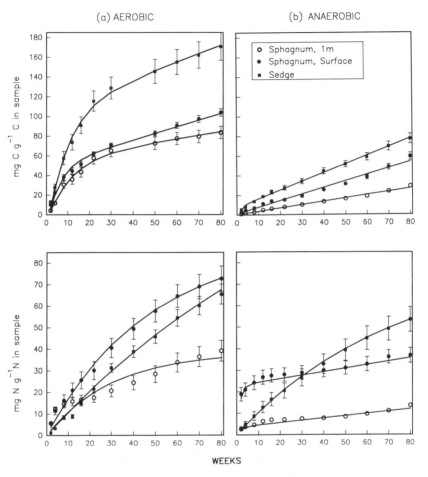

Fig. 15-2. (*a*) Cumulative and (*b*) anaerobic C and N mineralization from histosols (*Sphagnum* surface and 1 m) and mineral soils (sedge), as a proportion of total C or N in the samples. Lines represent modeled mineralization kinetics (see Table 15-2), points are measured data. Error bars are ± 1 SE (*n* = 10); no error bars are shown wehre they are smaller than the symbol.

under aerobic than under anaerobic conditions (Fig. 15–2). Cumulative N mineralization follows first-order kinetics (Eq. [1]) in all soils under aerobic conditions, and in the anaerobic sedge soil (Table 15–2), but not in the anaerobic *Sphagnum* peats. Note that the relative ranks of the anaerobic surface peat and sedge soil samples reversed after 30 wk of incubation. Short-term incubations, such as the 14-d anaerobic incubation of Waring and Bremner (1964), would not have resolved the long-term kinetics or indicated the rank reversal.

There were also significant interactions between soil type and aeration status. The sedge soil mineralized N at 1.4 to 4 times the rate of *Sphagnum* peats under anaerobic conditions. However, aeration caused a two- to threefold increase in N mineralization in the peats, compared with an insignificant response in the sedge soils (Fig. 15-2). This resulted in greater aerobic N mineralization in the surface *Sphagnum* peats than in the sedge soil.

Carbon Mineralization

Soil type, aeration status, and their interaction were all significant ($P < 0.01$) in determining C mineralization, as with N mineralization. The sensitivity of N and C mineralization to aeration was similar in the *Sphagnum* peats, but in the sedge soil C mineralization was much more sensitive than N mineralization to aeration status. The sedge soils had the greatest C mineralization under both aerobic and anaerobic conditions (Fig. 15-2). The linear decay kinetics manifested by the anaerobic *Sphagnum* peats (Fig. 15-2) are in contrast to the clearly two-stage kinetics seen in the aerobic incubations. Apparently, the organic matter fraction subject to rapid decay under aerobic conditions is not similarly labile under anaerobic conditions. Mean CO_2-C/CH_4-C ratios under anaerobic conditions had declined from more than 300 to less than 10 in the *Sphagnum* peats after 30 wk of incubation. The initial ratio of 10 in the sedge soil declined only slightly over the course of the incubation. The drop in CO_2-C/CH_4-C ratios is attributable largely to an increase in CH_4-C production, whereas total C mineralization increased less or declined. The increase in CH_4 as a C-mineralization end product may be related to a gradual increase in methanogen populations, or to changes in substrate availability as decomposition progressed, since nutrients were supplied and Eh changed only slightly over the course of the incubation. A Pearson's correlation coefficient of 0.56 was determined between Eh and the log of CH_4 emission rate, however, the relationship was not consistent over time.

SUMMARY AND CONCLUSIONS

Estimating C and nutrient release rates from wetland soils is essential to modeling possible shifts in trace gas composition of the atmosphere. Although long-term aerobic soil incubations have been used to quantify C and N dynamics, which are directly related to productivity in soils, previously no equivalent method existed to provide similar data for anaerobic soils. We developed an anaerobic long-term soil incubation method for measuring C and nutrient mineralization over an indefinite period. We used this method in conjunction with a proven aerobic incubation technique (Nadelhoffer, 1990) to assess long-term C and N mineralization kinetics in three soils in relation to environmental variables and substrate characteristics.

During 80-wk incubations, aerobic mineralization of N and C was up to three times higher than anaerobic mineralization, although the difference was much less in a sedge meadow soil than in surface and 1-m *Sphagnum*-derived peats. Carbon and N mineralization showed similar sensitivities to aeration status in the peats, but in the sedge soil C mineralization responded more strongly than N mineralization to aeration. Under anaerobic conditions, the CO_2-CH/CH_4-C ratio declined as decomposition advanced. Some soil differences, in particular with respect to anaerobic N mineralization, would have been misinterpreted had only short-term incubations been used. Aerobic and anaerobic C and N mineralization kinetics were essentially similar in the sedge soil, but not in the *Sphagnum* peats (Table 15-2). This implies that greater increases in decomposition rates in response to drainage can be expected from *Sphagnum* bogs than from sedge-type wetlands.

The parallel incubation technique presented here permits the simultaneous assessment of long-term soluble nutrient and C mineralization patterns under controlled aerobic and anaerobic conditions. The incubations provided useful comparisons of N and C mineralization dynamics in relation to aeration status and organic matter quality in three wetland soils. Although only N and C mineralization were considered in this instance, in subsequent experiments we have also measured P mineralization, and the technique could plausibly be used to quantify the mineralization dynamics of any water-soluble nutrient.

ACKNOWLEDGMENTS

This research was supported by a grant from the U.S. National Aeronautics and Space Administration (NASA/NAGW - 1828) to Carol Johnston and John Pastor, and a U.S. Dep. of Energy Global Change Distinguished Postdoctoral Fellowship to Scott Bridgham. The authors thank Brad Dewey and Anastasia Bamford for technical assistance.

REFERENCES

Binkley, D., and S.C. Hart. 1989. The components of nitrogen availability assessment in forest soils. Adv. Soil Sci. 10:57–112.

Blet-Charaudeau, C., J. Muller, and H. Laudelout. 1990. Kinetics of carbon dioxide evolution in relation to microbial biomass and temperature. Soil Sci. Soc. Am. J. 54:1324–1328.

Bonde, T.A., and T. Lindberg. 1988. Nitrogen mineralization kinetics in soil during long-term aerobic laboratory incubations: A case study. J. Environ. Qual. 17:414–417.

Bonde, T.A., and T. Rosswall. 1987. Seasonal variation of potentially mineralizable nitrogen in four cropping systems. Soil Sci. Soc. Am. J. 51:1502–1514.

Cassman, K.G., and D.N. Munns. 1980. Nitrogen mineralization as affected by soil moisture, temperature and depth. Soil Sci. Soc. Am. J. 44:1233–1237.

Faulkner, S.P., W.H. Patrick, Jr., and R.P. Gambrell. 1989. Field techniques for measuring wetland soil parameters. Soil Sci. Soc. Am. J. 53:883–890.

Gale, P.M., and J.T. Gilmour. 1988. Net mineralization of carbon and nitrogen under aerobic and anaerobic conditions. Soil Sci. Soc. Am. J. 52:1006–1010.

Mah, R.A., D.M. Ward, L. Baresi, and T.L. Glass. 1977. Biogenesis of methane. Ann. Rev. Microbiol. 31:309–341.

McClaugherty, C.A., J. Pastor, J.D. Aber, and J.M. Melillo. 1985. Forest litter decomposition in relation to soil nitrogen dynamics and litter quality. Ecology 66:266–275.

McGill, W.B., and C.V. Cole. 1981. Comparative aspects of cycling organic C, N, S and P through soil organic matter. Geoderma 26:267–286.

Moore, T.R., and R. Knowles. 1989. The influence of water table levels on methane and carbon dioxide emissions from peatland soils. Can. J. Soil Sci. 69:33–38.

Nadelhoffer, K.J. 1990. Microlysimeter for measuring nitrogen mineralization and microbial respiration in aerobic soil incubations. Soil Sci. Soc. Am. J. 54:411–415.

Powers, R.F., 1980. Mineralizable soil nitrogen as an index of N availability to forest trees. Soil Sci. Soc. Am. J. 44:1314–1320.

Reeburgh, W.S. 1983. Rates of biogeochemical processes in anoxic sediments. Ann. Rev. Earth Plant Sci. 11:269–298.

Schimel, D.S., W.J. Parton, T.G.F. Kittel, D.S. Ojima, and C.V. Cole. 1990. Grassland biogeochemistry: Links to atmospheric processes. Climatic Change 17:13–25.

Stanford, G., and S.J. Smith. 1972. Nitrogen mineralization potentials of soils. Soil Sci. Soc. Am. J. 36:465–472.

Waring, S.A., and J.M. Bremner. 1964. Ammonium production in soil under waterlogged conditions as an index of nitrogen availability. Nature (London) 201:951–952.

Yavitt, J.B., G.E. Lang, and R.K. Wieder. 1987. Control of carbon mineralization to CH_4 and CO_2 in anaerobic, *Sphagnum*-derived peat from Big Run Bog, West Virginia. Biogeochemistry 4:141–157.

16 Fungal Contributions to Soil Aggregation and Soil Quality

Neal S. Eash

Iowa State University
Ames, Iowa

Douglas L. Karlen and Timothy B. Parkin

USDA-ARS
National Soil Tilth Laboratory
Ames, Iowa

Increased public concern for improved environmental quality necessitates determination of the mechanisms important in soil structure formation to decrease runoff and erosion (Karlen et al., 1992). A prerequisite to evaluating soil quality is understanding factors influencing soil structure. Yoder (1937) stated, "Anyone who is conservation minded, and at the same time familiar with the close correlation between poor tilth and high erodibility, is inclined to the view that the possibility of improving soil structure should be investigated to the fullest extent." This statement remains relevant today.

Historically, soil *tilth* has holistically described soil in a conceptual way: ease of tillage, fitness as a seedbed, ability to support plant growth (SCSA, 1982) without recognizing the contributions of the soil biota. Increased interest in the biological contributions to soil tilth has recently resulted in the concept of soil health. A healthy soil is thought to be a biologically active soil with a large and diverse microbial biomass pool. Soil quality encompasses the concepts of soil tilth and soil health; a quality soil would exhibit characteristics of high health and high tilth.

Soil structure has been defined as the size and arrangement of particles and pores in soil (Oades, 1984). Soil structure is created when physical forces (drying, shrink–swell, freeze–thaw, root growth, animal movement, or compaction) mold the soil into aggregates (Paul & Clark, 1988). Soil structure controls the formation and destruction of soil organic matter, soil porosity, and water and solute movement through the soil profile (Elliott & Coleman, 1988).

The formation of soil structure results in aggregate formation. As aggregates become larger, the architecture, arrangement, and stability of the aggregates is important because these factors dictate pore size and continuity. Aggregate breakdown relates closely to infiltration and erodibility in the field (Coughlan et al., 1991; Yoder, 1937). If the soil aggregates are unstable during a rainfall event, decreased water infiltration results, due to partial blocking of the pores with aggregate particles (Lynch & Bragg, 1985). As the percentage of water-stable aggregates increase, water infiltration increases with concomitant decrease in erosional losses (Wilson & Browning, 1945). Therefore, aggregate size and stability control the soil pore-size distribution, water infiltration rates, and erosivity.

Numerous research studies indicate the importance of crop residues on promoting aggregation. Crop rotations with low organic matter inputs lead to loss of water-stable aggregates (Boyle et al., 1989; Oades, 1984). Residue management also affects the microbial community composition (Holland & Coleman, 1987; Doran, 1980). However, the actual mechanisms contributing to this loss in aggregation are not well defined.

Numerous writers have implicated soil microorganisms as important factors in soil aggregation and soil structure. A study by Martin and Waksman (1940) indicated that microorganisms are important in promoting soil structure and reducing erosional losses. Hierarchical models of biological contributions to soil structure and aggregation imply that bacteria and fungi promote aggregation in soil. A model developed by Tisdall and Oades (1982) suggests that the bacteria are important at one size-fraction and that fungi are important at another size fraction; however, data to support this model are not included in their paper. A study by Gupta and Germida (1988), which contrasted native and cultivated soil, indicated that the fungal biomass was important in the formation of soil aggregates. In their field study fungal biomass correlated with increases in aggregate mean weight diameter. Their results suggest that the reduced aggregate stability following cultivation can be explained by decreases in total microbial biomass coupled with significant loss of fungal biomass, but no controlled laboratory studies were performed to evaluate this hypothesis.

A deficiency exists in the literature delineating effects of fungi and residue on aggregating soil. The mechanism by which residue stabilizes soil is unclear. While clay–organic matter complexes may increase aggregation at one aggregate-size class, residue additions to soil also provide substrate for increased microbial activity. The purpose of this experiment was to evaluate the potential impact that a soil fungus may have on soil aggregation and to assess the *nonbiological* effects of residue on soil aggregation.

In this laboratory study, the interactions of residue and *Chaetomium* sp. impacting two soil materials of different textures were investigated. Aggregation was determined by measuring wet aggregate stability. Fungal growth was monitored using ergosterol and direct microscopic examination of hyphal length. Specifically, the objectives of this study were:

1. To determine the effect of alfalfa (*Medicago sativa* L.) residue, independent of biological activity, on soil aggregation
2. To determine if the addition of fungal inoculum into autoclaved soil increases soil aggregation.

MATERIALS AND METHODS

To accomplish the objectives of this laboratory experiment, a factorial design with two levels of fungi (inoculated and uninoculated), two levels of residue (alfalfa residue added and no crop residue added), two soils (Clarion loam and a silt loam parent material), and three replicates of each were used for a total of 24 jars.

Soil materials used in this experiment were basal loess parent material collected below an Ida silt loam (1.8-m depth; fine-silty, mixed, mesic, calcareous Typic Eudorthents) in western Iowa and a Clarion loam (0 to 15-cm depth; fine-loamy, mixed, mesic Typic Hapludolls) from the research farm near Ankeny, IA. These two soils were chosen based on their total C content and texture. The silt loam parent material consists of 78% silt. Because of the erosiveness of silt-size particles and the fact that western Iowa has some of the highest rates of soil loss in the world, the low C parent material was selected. The A horizon of soil mapped Clarion series was selected because it is very common in central Iowa and has a relatively high total C content (20.4 g kg^{-1}) in contrast with the silt loam parent material (1.4 g kg^{-1}). The soil material was air-dried and ground to pass a 2-mm sieve. A pressure plate apparatus was used to determine 33 kPa water contents.

Soil (400 g) was placed into 0.95-L (1 quart) Mason jars and capped with lids that contained a rubber stopper through which a Pasteur pipette was inserted. Glass wool was placed into the Pasteur pipette so that air could pass into the jar without contaminating the sample. All jars were autoclaved on 2 successive days for 30 min at 121 °C at 101.4 kPa pressure. Following the final autoclaving, all samples was aseptically brought to 33 kPa water content with sterile water. Sterilization of the samples was verified by sampling for CO_2, in the headspace of jars used as controls. Effect of alfalfa residue on soil aggregation was evaluated by adding 2 g of alfalfa that had been ground in a Wiley mill to pass a 2-mm screen. In treatments with alfalfa residue added, the residue was added to the soil and thoroughly mixed prior to autoclaving.

Samples were incubated at room temperature with normal fluorescent lighting for 7 d. Preliminary laboratory studies indicated that most of the aggregation occurs in the first 7 to 10 d of incubation. These results are similar to other published research studies evaluating biological contributions to soil aggregation (Molope et al., 1985; Molope, 1987).

Inoculum

A preliminary experiment using nine ergosterol–producing fungi indicated that *Chaetomium* sp. showed average response in terms of promotion of soil aggregation. *Chaetomium* sp. was grown in acidified potato dextrose broth for 7 d on a reciprocal shaker. Streptomycin sulfate (0.001 kg/L) and chloramphenicol (0.001 kg/L) were added to the cultures as bactericides. Prior to inoculating the jars, the cultures were examined for bacteria by direct microscopic counting. Cultures were placed into a sterile Waring blender and blended at high speed for 1 min. Immediately after the blender was stopped, 1 mL of the chopped fungus was added to the autoclaved jars that were to be inoculated. Ergosterol content and hyphal length was measured on the inoculum.

In this study, wet aggregate stability was used, because it is a measure of soil erodibility by water (Yoder, 1936). Following the incubation period, the jars were sampled as follows. The soil in the jar was removed, quartered, and sampled for water content. A nest of sieves (8, 4, 2, 1, 0.5, 0.25, and 0.125 mm) were used in determining aggregate stability (Yoder, 1936) on 200 g of moist soil. The samples were immediately immersed and sieved, and no prewetting occurred. Wet sieving was completed on a sieving machine similar to the machine illustrated in Low (1954). The soil samples were immersed in water and sieved for 3 min at 130 cycles per min. Results were calculated using aggregate mean weight diameter (MWD) (Kemper & Rosenau, 1986).

Ergosterol and Hyphal Length

Ergosterol analyses followed the method developed by Grant and West (1986). Development of an ergosterol assay by Seitz et al. (1977) and its modification by Grant and West (1986) provided a relatively fast assay for determining live fungal biomass by quantifying soil ergosterol content (West et al., 1987). Ergosterol is the predominant sterol of most fungi (Weete, 1973; Grant & West, 1986; Davis & LaMar, 1992). Ergosterol correlates with fungal surface area (West et al., 1987), hyphal length (Matcham et al., 1985), and is more sensitive than the chitin or extracellular laccase assays (Matcham et al., 1985; Seitz et al., 1979). Therefore, ergosterol may be an important indicator of changes in the fungal portion of the soil microbial biomass.

Hyphal length was determined by direct microscopic observation of diluted soil samples using the membrane filtration technique (Hansen et al., 1974). Soil (1 g) was added to 500 mL of distilled and filtered water, then blended (Osterizer Pulsematic 14, Milwaukee, WI) at high speed for 1 min. A 1-mL aliquot was removed while operating the blender at low speed and placed into a filtering apparatus. Black polycarbonate filters (25 mm diam., 0.4 μm) (Poretics Corp, Livermore, CA) were used with 25 mm polyester drain disks placed under the polycarbonate filters. Vacuum was applied to filter the sample. Calcofluor white M2R (Sigma) fluorescent stain (1 mL of 0.002 g mL^{-1} distilled and filtered water) was expelled through a 0.2-μm syringe filter into the filtering apparatus and allowed to stand for at least 1 min (West, 1988). Vacuum was applied to pull the stain through the filter. The filter was washed with five, 1-mL aliquots of water. The filters were removed, placed on glass slides, and placed onto a slightly warm hotplate for approximately 10 min to dry the filter prior to mounting the cover slips. Cover slips were mounted over the filters using immersion oil. Hyphal length was determined at 400X using epifluorescence microscopy and a Nikon UV-1A filter cube on a Nikon microscope (Nikon Inc., Melville, NY). Hyphal length was calculated using the grid line intersect method developed by Olson (1950).

Homogeneity of variance was analyzed with Box-Cox transformations (Box et al., 1978, p. 231–238). Each significant ($P < 0.05$) variable with heterogeneous variance was transformed with the appropriate Box-Cox transformation and analyzed with the appropriate analysis of variance. Least squares analysis of variance was used on treatments with missing samples. Treatment differences ($P < 0.05$) were examined with planned contrasts.

RESULTS AND DISCUSSION

Autoclaved soil was used to determine effects of residue alone on soil aggregation. In autoclaved soil, the addition of ground alfalfa residue did not significantly increase soil aggregation (Table 16–1). Table 16–1 indicates that wet MWD decreased significantly in the Clarion loam soil. These results indicate that residue, in the absence of an active microbial component, does not increase aggregation. This decrease in aggregation may be due to the residue physically interfering with the effective attachment of individual soil particles.

Ergosterol and hyphal length measurements confirm that fungal growth occurred following inoculation of the samples (Table 16–2). The silt loam parent material had increased ergosterol content and a doubling of hyphal length in the inoculated samples when compared with the uninoculated samples. These increases, however, are much smaller than the increases found in the Clarion soil material. Addition of the fungal inoculum resulted in a ninefold increase in hyphal length. Large increases in hyphal length following addition of the inoculum to the Clarion soil not amended with alfalfa were observed. Autoclaving the Clarion soil may have liberated soil organic C, which could serve as substrate for the fungal inoculum.

Alfalfa residue additions to the silt loam parent material resulted in an 11-fold increase in hyphal length and a similar increase in ergosterol content (Table 16–2). Alfalfa additions to the Clarion soil material did not significantly increase aggregate stability (Table 16–3).

Increases in either soil ergosterol content or hyphal length corresponded with increased wet MWD (Tables 16–2 and 16–3). Increases in wet MWD

Table 16–1. Effects of alfalfa residue on soil aggregation using autoclaved soil.[†]

Residue added	Wet WMD[‡]
Silt loam parent material	
None	0.02a*
Alfalfa	0.06a
Clarion loam	
None	3.32a
Alfalfa	2.78b

* Means with the same letter are not significantly different ($P > 0.05$). Comparisons are between treatments of the same texture.
[†] Data are means of three replicates.
[‡] Aggregate mean weight diameter.

Table 16-2. Fungal biomass as affected by residue or fungal inoculum addition.†

	No fungal inoculum added		Fungal inoculum added	
Soil	Ergosterol	Hyphal length	Ergosterol	Hyphal length
	mg kg^{-1}	m g^{-1}	mg kg^{-1}	m g^{-1}
Silt loam parent material				
Control	0.00a*	6.99a	0.79a	16.30a
Alfalfa	0.00a	20.46a	11.41b	182.89b
Clarion loam				
Control	0.07a	18.39a	1.54a	157.29a
Alfalfa	0.10a	29.56a	3.06a	233.82a

* Means with the same letter are not significantly different ($P > 0.05$). Comparisons are between treatments of the same texture within inoculum treatment.
‡ Data are means of three replicates.

Table 16-3. Effect of adding *Chaetomium* sp. and alfalfa residue on soil aggregation.†

		No residue added	Alfalfa residue added
Soil	Inoculated	Wet MWD‡	Wet MWD
		mm	
Silt loam parent material	No	0.02a*	0.06a
Silt loam parent material	Yes	0.22b	1.88b
Clarion loam	No	3.32a	2.78a
Clarion loam	Yes	4.43b	4.04b

* Means with the same letter are not significantly different ($P > 0.05$). Comparisons are between soil materials of the texture and residue treatment.
† Data are means of three replicates.
‡ Aggregate mean weight diameter.

were significantly correlated with increased hyphal length and soil ergosterol content (N.S. Eash et al., 1992, unpublished data). Inoculation of the autoclaved silt loam parent material resulted in a 10-fold increase in MWD of the soil aggregates. Aggregate MWD of the well-aggregated Clarion soil material significantly increased with the addition of the fungal inoculum.

The samples inoculated with *Chaetomium* sp. had significantly higher aggregate mean weight diameters as measured by wet sieving when compared with the autoclaved, uninoculated samples (Table 16-3). Addition of fungal inoculum to soil and the resultant growth of the fungi, as evidenced by increased ergosterol content and hyphal length, indicates that fungi are important mechanisms in aggregation.

Results from this experiment indicate that enhanced aggregation in soil following residue additions, as reported in other studies (Wilson & Browning, 1945; Oades, 1984), is not due to residue per se; rather, it is probably caused by greater fungal activity. Residue cover on the soil surface absorbs the energy of falling raindrops, but, perhaps more importantly, provides substrate for the soil fungi. By providing substrate, increased growth of fungal hyphae can physically stabilize soil particles into larger aggregates.

If soil aggregation cause-and-effect relationships were implied from past research data, tillage, rotation, or residue cover were often the likely reasons cited. This research does not dispute past observations; instead, it demonstrates quantitatively that soil fungi play a major role in soil aggregation. As we strive for a more sustainable future for our agricultural systems, the interactions of the fungi and soil management procedures are an important consideration.

REFERENCES

Box, G.E.P., W.G. Hunter, and J.S. Hunter. 1978. Statistics for experimenters, an introduction to design, data analysis, and model building. John Wiley & Sons, New York.

Boyle, M., W.T. Frankenburger, Jr., and L.H. Stolzy. 1989. The influence of organic matter on soil aggregation and water infiltration. J. Prod. Agric. 2:290-299.

Coughlan, K.J., D. McGarry, R.J. Loch, B. Bridge, and G.D. Smith. 1991. The measurement of soil structure—some practical initiatives. Aust. J. Soil Res. 29:869-889.

Davis, M.W., and R.T. LaMar. 1992. Evaluation of methods to extract ergosterol for quantitation of soil fungal biomass. Soil Biol. Biochem. 24:189-198.

Doran, J.W. 1980. Soil microbial and biochemical changes associated with reduced tillage. Soil Sci. Soc. Am. J. 44:765-771.

Elliott, E.T., and D.C. Coleman. 1988. Let the soil work for us. Ecol. Bul. 39:23-32.

Grant, W.D., and A.W. West. 1986. Measurement of ergosterol, diaminopimelic acid and glucosamine in soil: Evaluation as indicators of microbial biomass. J. Microbiol. Methods 6:47-53.

Gupta, V.V.S.R., and J.J. Germida. 1988. Distribution of microbial biomass and tis activity in different soil aggregate size classes as affected by cultivation. Soil Biol. Biochem. 20:777-786.

Hansen, J.F., T.F. Thingstad, and J. Godsoyr. 1974. Evaluation of hyphal lengths and fungal biomass in soil by a membrane filter technique. Oikos 25:102-107.

Holland, E.A., and D.C. Coleman. 1987. Litter placement effects on microbial and organic matter dynamics in an agroecosystem. Ecology 68:425-433.

Karlen, D.L., N.S. Eash, and P.W. Unger. 1992. Soil and crop management effects on soil quality indicators. Am. J. Altern. Agric. 7:48-55.

Kemper, W.D., and R.C. Rosenau. 1986. Aggregate stability and size distribution. p. 432-433. In A. Klute (ed.) Methods of soil analysis. Part 1. 2nd ed. Agron. Monogr. 9. ASA and SSSA, Madison, WI.

Low, A.J. 1954. The study of soil structure in the field and in the laboratory. J. Soil Sci. 5:57-78.

Lynch, J.M., and E. Bragg. 1985. Microorganisms and soil aggregate stability. Adv. Soil Sci. 2:133-171.

Martin, J.P., and S.A. Waksman. 1940. Influence of microorganisms on soil aggregation and erosion. Soil Sci. 50:29-47.

Matcham, S.E., B.R. Jordan, and D.A. Wood. 1985. Estimation of fungal biomass in a solid substrate by three independent methods. Appl. Microbiol. Biotechnol. 21:108-112.

Molope, M.B. 1987. Soil aggregate stability: The contribution of biological and physical processes. S. Afr. J. Plant Soil 4:121-126.

Molope, M.B., I.C. Grieve, and E.R. Page. 1985. Thixotropic changes in the stability of molded soil aggregates. Soil Sci. Soc. Am. J. 49:979-983.

Oades, J.M. 1984. Soil organic matter and structural stability: Mechanisms and implications for management. Plant Soil 76:319-337.

Olson, F.C.W. 1950. Quantitative estimates of filamentous algae. Trans. Am. Microsc. Soc. 69:272-279.

Paul, E.A., and F.E. Clark. 1988. Soil microbiology and biochemistry. Academic Press, New York.

Seitz, L.M., H.E. Mohr, R. Burroughs, and D.B. Sauer. 1977. Ergosterol as an indicator of fungal invasion in grain. Cereal Chem. 54:1207-1217.

Seitz, L.M., D.B. Sauer, R. Burroughs, H.E. Mohr, and J.D. Hubbard. 1979. Ergosterol as a measure of fungal growth. Phytopathology 19:1202–1203.

Soil Conservation Society of America. 1982. Resource conservation glossary. SCSA, Ankeny, IA.

Tisdall, J.M., and J.M. Oades. 1982. Organic matter and water-stable aggregates in soils. J. Soil Sci. 33:141–163.

Weete, J.D. 1973. Sterols of the fungi: Distribution and biosynthesis. Phytochemistry 12:1843–1864.

West, A.W. 1988. Specimen preparation, stain type, and extraction and observation of soil mycelial lengths and volumes by light microscopy. Biol. Fertil. Soils 7:88–94.

West, A.W., W.D. Grant, and G.P. Sparling. 1987. Use of ergosterol, diaminopimelic acid and glucosamine contents of soils to monitor changes in microbial populations. Soil Biol. Biochem. 19:607–612.

Wilson, H.A., and G.M. Browning. 1945. Soil aggregation, yields, runoff, and erosion as affected by cropping systems. Soil Sci. Soc. Am. Proc. 10:51–57.

Yoder, R.E. 1936. A direct method of aggregate analysis of soils and a study of the physical nature of erosion losses. J. Am. Soc. Agron. 28:337–351.

Yoder, R.E. 1937. The significance of soil structure in relation to the tilth problem. Soil Sci. Soc. Am. Proc. 2:21–33.

17 Microbial Biomass as an Indicator of Soil Quality: Effects of Long-Term Management and Recent Soil Amendments

Mary F. Fauci and Richard P. Dick

Oregon State University
Corvallis, Oregon

Crop rotations, residue management, fertilization, cultural, and other management practices can significantly affect soil quality by changing soil physical, chemical, and biological parameters. When management practices remain constant, soil ecosystems approach an equilibrium that reflects the soil quality.

Soil quality indicators must be capable of discriminating between the effects of various management practices on the soil resource. Soil organic matter, measured as total organic C, is one of the properties most widely used to describe soil quality. Soil microbial biomass (MB) is closely related to organic matter content. Because the C and N in the MB (MBC and MBN, respectively) turn over rapidly and reflect changes in management practices long before changes in total soil C or N are detectable (Powlson et al., 1987), MB has potential as a soil quality indicator.

Since 1931, the Residue Utilization Plots (RUPs) at the Columbia Basin Research Center, Pendleton, OR, have been cropped to a wheat (*Triticum aestivum* L.)–fallow system and managed with organic or inorganic N sources. As a result of different long-term N treatments, soil quality as measured by physical, chemical, and biological properties varies widely (Collins et al., 1992; Dick et al., 1988; Rasmussen et al., 1980, 1989). The objective of this chapter is to report on the potential of using MB as a soil quality index to discriminate between long- and short-term manipulations of organic and inorganic N amendments in soils.

MATERIALS AND METHODS

The RUPs established in 1931 at the Columbia Basin Research Center, Pendleton, OR, are in a wheat–fallow crop rotation. Treatments include different N fertilizer rates and sources of organic material. Soil was collected from the fallow phase of the rotation in November 1989 for a greenhouse study. In the greenhouse, four crops of corn (*Zea mays* L.) were grown over a period of 306 d in pots containing 2 kg soil. The experimental design of the greenhouse study was a completely randomized block with three replications. Treatments were arranged as a complete factorial (4 × 4 × 4). The three factors were RUP soil and greenhouse organic-N and inorganic-N amendments. Soil came from the following four RUPs: (i) control, no added N; (ii) inorganic N, 90 kg N ha^{-1} as NH_4NO_3 applied biennially prior to seeding; (iii) beef manure, 22 Mg (dry wt.) manure ha^{-1} applied each fallow year; and (iv) pea (*Pisum sativum* L.) vine, 2 Mg (dry wt.) pea crop residue ha^{-1} applied each fallow year. All treatments had wheat straw incorporated. Organic amendments added in the greenhouse were pea vine, beef manure, poultry manure, or control (no organic amendment added). Inorganic N was added as NH_4NO_3 at four rates ranging from 0 to 400 mg N 2 kg^{-1} soil per crop.

Organic amendments were mixed into the soil on an equal N basis (1 g total N 2 kg^{-1} soil) before cropping began and again after the second crop was harvested. An incubation period followed both additions of organic amendments. Half of the inorganic N was surface-applied at planting and the other half 21 d after planting. The high (400 mg N 2 kg^{-1} soil) and zero N rates remained constant for each crop. The two intermediate rates decreased during the experiment. Further description of the experiment was reported by Fauci and Dick (1994).

Soil MB analysis was performed prior to the greenhouse experiment (Day 0), after the soil was incubated with the organic residues but before any crop was grown (Day 87), after the second harvest (Day 164), and after the fourth harvest (Day 306). Microbial biomass C and MBN were determined by the chloroform–fumigation incubation method of Jenkinson and Powlson (1976) on a 12-g (fresh wt.) soil sample. The amount of CO_2 evolved during the 10-d incubation following fumigation was determined on a thermal conductivity gas chromatograph. After CO_2 sampling, NH_4^+-N was extracted with 50 mL of 2 *M* KCl and quantified with automated colorimetric analysis (Alpkem, Clakamas, OR). The MBC and MBN were calculated with the following formulas: MBC = CO_2-Cf/0.41 (Voroney & Paul, 1984), MBN = (NH_4^+-Nf − NH_4^+-Nuf)/0.68 (Shen et al., 1984), where f and uf denote fumigated and unfumigated samples, respectively.

All results are expressed on a per kg oven dry (105 °C, 24 h) weight basis. The data were analyzed by standard ANOVA techniques for a randomized complete-block design with SAS statistical software package (SAS Inst., Cary, NC).

RESULTS AND DISCUSSION

Effect of Long-Term Management

Results showed that MB discriminated between long-term effects. In the absence of additional amendments, soil in the manure RUP had greater MBC

Table 17-1. Effects of organic and inorganic N treatments on microbial biomass C in soil from the long-term Residue Utilization Plots, Pendleton, OR.

Greenhouse treatments	Long-term treatments			
	Manure	Pea vine	N	Control
	mg C kg^{-1} soil			
Control $_{Day\ 87\dagger}$	145a*	104b	73c	74c
Control $_{Day\ 306\dagger}$	165a	106b	113b	90b
Pea vine‡	536a	512b	521a	440b
Beef manure‡	354a	341a	304a	279a
Poultry manure‡	264a	162a	190a	185a
Nitrogen$_{400}$§	155a	103ab	125ab	72b
Nitrogen$_{800}$§	144a	98b	73b	86b
Nitrogen$_{1600}$§	116a	77ab	68ab	66b

* Means within each row rollowed by the same letter are not significantly different at $P = 0.05$ according to the least significant difference test.
† These controls did not receive N (inorganic or organic) in the greenhouse. Sampled prior to cropping (Day 87) or after Crop 4 harvest (Day 306).
‡ These treatments received no inorganic N amendments in the greenhouse. Sampled on Day 306.
§ These treatments received no organic amendments in the greenhouse. Subscript numnber is the cumulative mg N applied per plot to the four crops. Sampled on Day 306.

(Table 17-1) and MBN (data not shown) than soil from any of the other RUP treatments. Long-term pea vine additions significantly increased MBN (data not shown) and tended to increase MBC (Table 17-1) relative to the control. This is consistent with other studies where MB differentiated between long-term soil amendments that varied in C quantity and N concentration (Collins et al., 1992; Schnürer et al., 1985). Microbial biomass can be increased directly by soil amendments or indirectly by the return of more residues, created as a result of improved plant nutrition (Insam et al., 1991).

In the RUPs, N treatment increased straw-C input by 26% in the field, but did not change the MB relative to the control. Decreased soil pH, in the long-term N treatment, probably counterbalanced the beneficial effect of increased C input on MB (Collins et al., 1992). Microbial biomass is an integrative measure of the soil environment, and thereby may minimize the number of indicator measurements needed to predict changes in soil quality.

Long-term crop yields from the RUPs indicate that the inorganic N and manure plots soils produce similar yields in the field (Rasmussen et al., 1989). Inorganic N can produce similar yields to organic inputs; however, it does not maintain soil organic matter or MB as well. Soil quality and soil productivity are not necessarily synonymous.

Effect of Soil Amendments

Recent addition of organic amendments had a large effect (80–400% increase over unamended soils) on the MB (Table 17-1). However, differences resulting from long-term field treatments were still evident. Soil from the long-term manure treatment still had higher MBC than soil from the other RUPs after 306 d, although this trend was not statistically significantly. Similar results were found for MBN (data not shown). Microbial biomass reflects past inputs over many years, but can be strongly influenced by a single organic input.

Differences in MB response to the organic amendments was probably the result of substrate availability or amount of C input. Residues were added on an equal N basis. The pea vine and beef manure added approximately the same amount of C (21 and 24 g C 2 kg^{-1} soil, respectively). Lignin concentration in the beef manure was four times greater than that in the pea vine (28 and 6% lignin, respectively). Because lignin is relatively resistant to microbial degradation, less C would be available for MB incorporation from the beef manure. The poultry manure added approximately one-third less C (8 g C 2 kg^{-1} soil) than either beef manure or pea vine. With a C/N ratio of 8.3, it is likely that C and not N was limiting the MB in the poultry manure treatment. These results indicate that MB is sensitive in discriminating between the effects of recent amendments in relation to their chemical composition.

Plant response in the soil amended with these organic materials was different than MB response. In the absence of inorganic N, total dry matter yield in the pea vine, poultry manure, beef manure, and control treatments was 32.6, 30.5, 7.4, and 9.4 g 2 kg^{-1} soil, averaged over long-term management treatments (Fauci & Dick, 1994). Although MB indicated changes in soil quality, this did not correspond to a related change in plant productivity. This is consistent with the results reported above on RUPs field crop productivity and MB.

The N fertilizer added in the greenhouse had only a significant effect on the MB when measured after Crop 4 harvest. When Crop 4 was grown, only the high N rate received any inorganic N. The MBN was 35% higher in the high N treatment than the lower N rates, which were not significantly different from one another (data not shown). Nitrogen fertilizer had no significant effect on MBC. Collins et al. (1992) also found N fertilization increased N concentration in MBN without affecting MBC in the RUPs and other plots located at the Columbia Basin Research Center. A few months after N fertilization, MBN had returned to levels similar to those measured prior to fertilization (Collins et al., 1992). Inorganic N causes a rapid, temporary increase MBN, but not in MBC. This transient effect may indicate that MBC is a better index of biological change than MBN.

There were rapid temporal effects from the organic amendments added in this study (Fig. 17–1). However, MB remained fairly constant in the soil that did not receive organic amendments (control). Since all crop residues were removed in the greenhouse, lack of C inputs in the control may have contributed to the initial decrease in MBC.

Perspectives

Microbial biomass C was sensitive in discriminating between long- and short-term soil amendments, including the subtle effects of substrate chemical composition. Microbial biomass C, unlike MBN, was unaffected by the

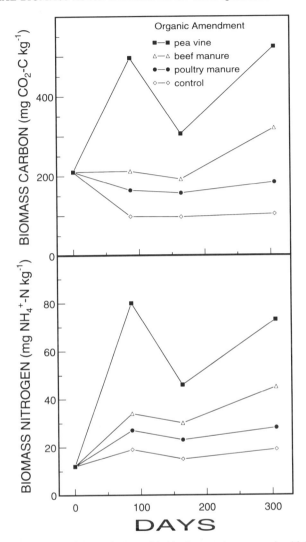

Fig. 17-1. Effect of organic amendments added in the greenhouse on microbial biomass C and N (no significant interactions; therefore, averaged over long-term and inorganic N treatments; $n = 48$).

direct effects of inorganic N. This may be an advantage for MBC as an index of soil biology, particularly for agricultural systems where the N that is routinely applied to soils would confound interpretations of MBN measurements.

A relatively rapid response to organic amendments did occur for MBC, which suggests it could be useful in identifying positive soil management effects on a temporal basis. For practical applications, MBC is attractive, because the procedure is relatively simple, unlike many other soil biological measurements. A limitation for MBC, however, is that recent short-term amendments can overshadow long-term effects and may not adequately reflect past management nor predict future changes in soil quality. To diminish this effect, it is important to avoid sampling after recent soil disturbances and standardize sampling during a similar phase of the cropping cycle to minimize seasonal, environmental, or human disturbances.

Microbial biomass is not a universal index and presently could only be useful in comparative situations (temporal or side-by-side treatments). This is supported by the results of this study where soil amendments caused wide fluctuations in MB, making a single MB measurement difficult to interpret. Furthermore, it was not related to crop productivity, indicating that MB may best be used as a component of an index that uses several key chemical and physical parameters in determining soil quality in relation to plant growth.

REFERENCES

Collins, H.P., P.E. Rasmussen, and C.L. Douglas. 1992. Crop rotation and residue management effects on soil carbon and microbial biomass dynamics. Soil Sci. Soc. Am. J. 56:783–788.

Dick, R.P., P.E. Rasmussen, and E.A. Kerle. 1988. Influence of long-term residue management on soil enzyme activities in relation to soil chemical properties of a wheat-fallow system. Biol. Fertil. Soils 6:159–164.

Fauci, M.F., and R.P. Dick. 1994. Plant response to organic amendments and decreasing inorganic nitrogen rates in soils from a long-term experiment. Soil Sci. Soc. Am. J. 58:(in press).

Jenkinson, D.S., and D.S. Powlson. 1976. The effect of biocidal treatments on metabolism in soil: V. A method for measuring soil biomass. Soil Biol. Biochem. 8:209–213.

Insam, H., C.C. Mitchell, and J.F. Dormaar. 1991. Relationship of soil microbial biomass and activity with fertilization practice and crop yield of three ultisols. Soil Biol. Biochem. 23:459–464.

Powlson, D.S., P.C. Brookes, and B.T. Christensen. 1987. Measurement of the soil microbial biomass provides an early indication of changes in total soil organic matter due to straw incorporation. Soil Biol. Biochem. 19:159–164.

Rasmussen, P.E., R.R. Allmaras, C.R. Rohde, and N.C. Roager. 1980. Crop residue influences on soil carbon and nitrogen in a wheat-fallow system. Soil Sci. Soc. Am. J. 44:596–600.

Rasmussen, P.E., H.P. Collins, and R.W. Smiley. 1989. Long-term management effects on soil productivity and crop yield in semi-arid regions of eastern Oregon. USDA-ARS Stn. Bull. Rep. 675.

Schnürer, J., M. Clarholm, and T. Rosswall. 1985. Microbial biomass and activity in agricultural soil with different organic matter contents. Soil Biol. Biochem. 17:611–618.

Shen, S.M., G. pruden, and D.S. Jenkinson. 1984. Mineralization and immobilization of nitrogen in fumigated soil and the measurement of microbial biomass nitrogen. Soil Biol. Biochem. 16:437–444.

Voroney, R.P., and E.A. Paul. 1984. Determination of k_C and k_N in situ for calibration of the chloroform fumigation–incubation method. Soil Biol. Biochem. 16:9–14.

18   The Response of Nematode Trophic Groups to Organic and Inorganic Nutrient Inputs in Agroecosystems[1]

Patrick J. Bohlen and Clive A. Edwards

Ohio State University
Columbus, Ohio

Free-living soil nematodes have an important role in the biotic regulation of soil nutrient cycling processes. Bacterivorous and fungivorous nematodes, through their feeding and excretory activities, increase the turnover and mineralization of nutrients that have been immobilized by the soil microflora (Yeates & Coleman, 1982; Coleman et al., 1984; Ingham et al., 1985; Freckman, 1988). The enhanced mineralization brought about by nematode feeding can increase the availability of nutrients for plant growth and the rate of turnover of microbial populations in soil (Clarholm et al., 1981; Ingham et al., 1985).

Nematodes have several characteristics that make them useful as a biological index of soil quality (Wasilewska, 1979; Bongers, 1990; Neher, 1992; Bernard, 1992; Stork & Eggleton, 1992). They are ubiquitous and have relatively short generation times and populations that respond rapidly to soil amendments and environmental variables (Bongers, 1990). Additionally, they are relatively easy to extract from soil. Major trophic categories of nematodes can be distinguished based on the known feeding habits of recognizable genera and gross morphology of their mouthparts and esophagus (Yeates, 1971; Niblack & Bernard, 1985). In addition, there is evidence that the relative proportions of fungivorous and bacterivorous nematodes may reflect the relative functional importance of fungi and bacteria to total microbial activity (Golebiowska & Ryszowski, 1977; Clarholm et al., 1981; Sohlenius & Bostrom, 1984; Beare et al., 1992).

Free-living nematode communities have not been studied widely in agricultural systems where most research has focused on economically important plant parasitic nematodes. The effect of organic amendments on nematode populations has been directed mainly toward using organic amend-

[1]This research was supported by the National Science Foundation Ecosystem Studies Program, Grant no. DEB-9020461.

ments to reduce plant parasitic nematode populations (Mankau, 1980; Freckman & Caswell, 1985; Freckman, 1988). Nematode communities, however, have been studied in tree nurseries (Niblack & Bernard, 1985), no-till systems (Baird & Bernard, 1984; Parmelee & Alston, 1986; Beare et al., 1992), and in Swedish perennial and annual farming systems (Sohlenius et al., 1987; Bostrom & Sohlenius, 1986). Changes in nematode community structure due to plowing of pastures (Sohlenius & Sandor, 1989) and during the course of decomposition of various plant residues (Wasilewska et al., 1981; Yeates & Coleman, 1982; Wasilewska & Bienkowska, 1985; Sohlenius & Bostrom, 1984) have shown that the resource quality of organic inputs has an influence on the trophic structure of nematode communities in agroecosystems (Freckman, 1988).

There is a need to understand the responses and dynamics of nematode communities to a wide array of management practices. More emphasis is being placed on sustainable management of agroecosystems, including more widespread use of more varied rotations, reduced or conservation tillage, cover crops, and organic amendments. It is necessary to understand how soil ecological processes are influenced by changes in utilization and placement of organic matter and other inputs. Trophic analysis of nematode communities, by itself and in conjunction with other ecosystem measurements, is a useful tool for studying biological responses to various inputs and practices. In this investigation we examined the short-term response of free-living nematode communities to inorganic fertilizer, animal manure, and plant residues added as nutrient inputs in corn (*Zea mays* L.) agroecosystems and laboratory microcosms.

MATERIALS AND METHODS

Field Study

This experiment was located at the Ohio Agricultural Research and Development Center in Wayne County, Ohio (41°N, 82°W). Mean monthly temperatures in the area range from -4.8°C in January to 21.2°C in July. Mean annual precipitation is 905 mm yr^{-1} and is fairly evenly distributed throughout the year. The site is a relatively flat, homogeneous area on a fine, mixed, mesic Fragiudalf soil of the Canfield series (Luvisol). Canfield soils are deep, gently sloping, moderately- to well-drained silt loam soils on uplands, with a relative impermeable fragipan at a depth of 40 to 75 cm. These soils represent a major agricultural soil type in the region. The initial total C and N contents at our site were 1.86 and 0.16%, respectively. Soil texture was 13.5% sand, 73.7% silt, and 12.8% clay. Soil pH was 8.6 and the CEC was 10 cmol$_C$ kg^{-1}. The field site was planted in corn until 1987 and in alfalfa (*Medicago sativa* L.) from 1987 to 1991, and received regular applications of liquid cow manure to both corn and alfalfa prior to the establishment of our experiment.

In spring 1991, 12 field plots (20 by 30 m) were laid out in a randomized block design with four replicates of each nutrient treatment. Each plot was assigned one of the following nutrient addition treatments at a rate of 100 kg N ha^{-1}: (i) NH$_4$NO$_3$ fertilizer, (ii) legume-cover crop, or (iii) cow manure. In 1991, all plots were planted to soybean [*Glycine max* (L.) Merr.] and no nutrients were added. In fall 1991, a rye (*Secale cereale* L.)/hairy vetch (*Vicia villosa* Roth) cover-crop was planted in the legume treatment plots. In spring 1992, corn was planted following incorporation of the cover-crop and addition of either NH$_4$NO$_3$ fertilizer or manure.

Nematode populations were sampled on 11 May, 9 June, 14 July, and 27 August in the 1992 growing season. Eighteen soil cores (5 cm diam. by 15 cm deep) were taken from each plot, nine within crop rows and nine between rows, at randomly selected locations. The cores were pooled and nematodes were extracted in Baermann funnels from two 20-g (wet wt.) subsamples (McSorley & Walter, 1991). Nematodes were separated into the following four trophic groups, based primarily on the morphology of the stoma and esophagus and known feeding habits of recognizable groups: fungivores, bacterivores, plant parasites, or omnivore–predators (Yeates, 1971; e.g., Parmelee & Alston, 1986).

Laboratory Microcosm Study

Microcosms consisted of 16-L cylindrical plastic containers (28 cm diam. by 38 cm height) filled with coarsely sieved (7-mm mesh) field-collected soil. The soil was collected from the field site on 12 Apr. 1991 and stored covered with a plastic sheet until processed. The soil was packed to approximate field bulk density (1.2 Mg m^{-3}). The bottoms of the microcosms had drainage holes and a 2-cm bed of small polyethylene beads. After the soil was added, microcosms were saturated thoroughly with distilled water and were frozen at $-25\,°C$ for 72 h to eliminate soil macroinvertebrates, particularly earthworm cocoons. Freezing was used by Huhta and Setala (1990) in another microcosm study and did not eliminate nematodes, although populations of bacterivores increased dramatically after freezing and those of predators were apparently reduced. A complete analysis of the effects of freezing on specific nematode groups was not done. Microcosms were maintained at room temperature under a 14 h light/10 h dark light regime. Each microcosm was allowed to equilibrate for 3 mo before nutrient amendments were added and received 2 cm of water biweekly during the study.

Nutrient amendments were added to the microcosms in quantities equivalent to 150 kg N ha^{-1} on 7 Jan. 1992. The three nutrient amendments were: (i) fresh straw-packed dairy cow manure; (ii) airdried, fresh hairy vetch residue, chopped in a food processor and moistened prior to addition; and (iii) granular ammonium nitrate. The nutrient amendments were mixed thoroughly with the top 5 cm of microcosm soil.

Nematode populations in the microcosms were sampled before adding nutrient inputs and at 4, 8, and 16 wk following additions. Three soil cores (2 by 15 cm) were taken from each microcosm on each sampling date. The cores were pooled and nematodes were extracted from two 20-g subsamples using Baermann funnels. Nematodes were classified into major trophic groups as previously described. Sampling holes were filled with soil taken from identical treatments that had been set up for this purpose.

Nematode data was analyzed with a General Linear Models Procedure using log-transformed data [ln $(x + 1)$] followed by Tukey's studentized range test (SAS Inst., 1985).

RESULTS

In the field study, addition of organic inputs led to significant increases in populations of fungivorous and bacterivorous nematode groups (Fig. 18–1).

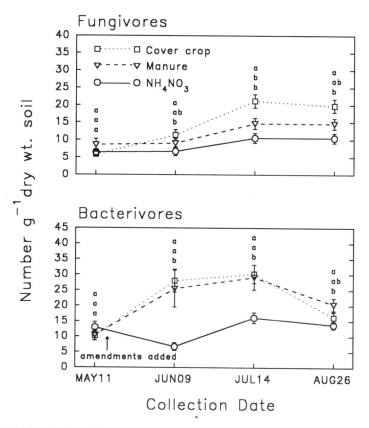

Fig. 18–1. Populations of fungivorous and bacterivorous nematodes (mean ± SE) in response to inputs of hairy vetch/rye cover crop, cow manure, and NH_4NO_3 fertilizer in a corn agroecosystem. Significant differences are indicated by different letters ($P = 0.05$).

In July, populations of fungivorous nematodes were significantly higher in the cover crop treatment than in the other two treatments. Bacterivores increased in number more rapidly than did fungivores following the addition of organic amendments. Population densities of bacterivores were similar under both the legume cover crop and animal manure treatments and were significantly higher than in the NH_4NO_3 treatment in June and July. Bacterivore populations were also significantly higher in the manure than in the NH_4NO_3 treatment in August (Fig. 18-1). Populations of bacterivores decreased slightly following the addition of NH_4NO_3 (Fig. 18-1, June) but returned to their preaddition levels by July. Except for this slight reduction, populations of both fungivorous and bacterivorous nematodes did not change significantly over the course of the growing season in the NH_4NO_3 treatment (Fig. 18-1).

The responses of populations of the nematode groups to nutrient inputs in the microcosm study were very similar to their responses in the field study, although the initial nematode populations were lower in the microcosm than in the field (Fig. 18-2). Fungivore populations increased signifi-

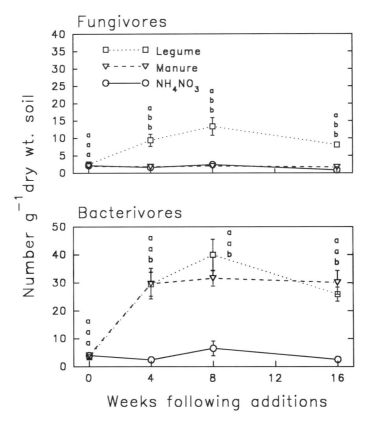

Fig. 18-2. Population of fungivorous and bacterivorous nematodes (mean ± SE) in response to hairy vetch residue (legume), cow manure, and NH_4NO_3 fertilizer in laboratory microcosms. Significant differences are indicated by different letters ($P = 0.05$).

Table 18-1. Nematode population density as affected by organic and inorganic nutrient inputs.

Nematode group	Nematode numbers (mean ± SE) summed over four dates		
	Legume	Manure	NH$_4$NO$_3$
	——— no. kg^{-1} dry wt. soil (x10^{-3}) ———		
Field study			
Total	152.03 ± 15.85	139.32 ± 19.44	102.87 ± 11.83
Fungivore	58.16 ± 6.36	46.77 ± 6.35	34.11 ± 4.88
Bacterivore	84.67 ± 10.61	85.45 ± 15.02	56.76 ± 7.06
Plant parasitic	7.56 ± 1.47	5.62 ± 1.34	10.69 ± 2.34
Omnivore–predator	1.60 ± 0.28	1.35 ± 0.41	1.35 ± 0.41
Bacterivore/Fungivore ratio	1.45	1.83	1.66
Microcosm study			
Total	138.90 ± 17.27	109.46 ± 13.16	29.87 ± 4.66
Fungivore	33.40 ± 5.55	7.52 ± 1.04	7.08 ± 1.25
Bacterivore	98.86 ± 12.75	95.51 ± 12.76	11.32 ± 3.58
Plant parasitic	4.83 ± 0.83	5.05 ± 0.53	6.98 ± 0.77
Omnivore–predator	1.79 ± 0.43	1.71 ± 0.33	0.43 ± 0.23
Bactivore/Fungivore ratio	2.96	12.7	1.60

cantly in response to the legume treatment. Although fungivore populations were slightly lower in the microcosm than in the field, in both experiments the overall population response of fungivores to the legume or cover crop treatment was similar. There were no significant differences in populations of fungivores between the manure and NH$_4$NO$_3$ treatments for the duration of the microcosm experiment. In these treatments, fungivore populations were very low and did not change significantly over the 16-wk term of the experiment.

As in the field study, numbers of bacterivores in the microcosms increased rapidly in response to organic amendments. Populations of bacterivorous nematodes were nearly identical in the legume and animal manure treatments on all sampling dates and were significantly higher than in the NH$_4$NO$_3$ treatment. Bacterivore population levels in the microcosms were very similar to the levels observed in the field (cf. Fig. 18-1 and 18-2).

Populations of plant parasitic and omnivore–predator nematodes were an order of magnitude lower than those of fungivores and bacterivores (Table 18-1). In both the field and microcosm studies, there was a slightly greater number of plant parasitic nematodes in the NH$_4$NO$_3$ treatment than in the other two treatments, and there were slightly more omnivores and predators in the legume treatment.

DISCUSSION

The trophic structure of free-living nematodes communities was significantly affected by the addition of organic nutrient amendments in both field and laboratory studies. Each type of organic input (cover crop residue or animal manure) elicited a characteristic response by populations of the microbivorous groups. Analysis of the bacterivore/fungivore ratios revealed that the bacterivores were relatively more abundant in the microcosms compared with the field, especially in the manure treatment (Table 18-1).

The greater proportion of bacterivores in the manure treatment relative to the legume treatment (Table 18-1) may reflect differences in the resource quality of these two organic inputs. The decomposition of cow manure is dominated by bacteria (Alejnikova et al., 1975). Sudhaus et al. (1988) studied the nematode species associated with manure decomposing in soil during the first 24 d of decomposition and found that nearly 100% of identified species were bacterivores. This predominance of bacterivores is consistent with the results of our microcosm study in which additions of manure increased populations of bacterivores significantly, but had no influence on fungivore populations over a period of 112 d (Fig. 18-2).

Plant residues contain a greater percentage of compounds that are recalcitrant to decomposition and are thus more likely to encourage greater fungal growth than animal manure. The greater fungal growth encouraged by plant residues may support larger populations of fungivorous nematodes (Clarholm et al., 1981; Sohlenius & Bostrom, 1984; Beare et al., 1992). Broder and Wagner (1988) compared microbial populations on corn, wheat (*Triticum aestivum* L.), and soybean residues. They found that the residue with the highest proportion of soluble components, soybean, had the highest bacterial and lowest fungal colony-forming units. Manure has a high proportion of soluble components, which may explain why bacteria predominate in manure decomposition.

In contrast with what occurred in the microcosms, fungivore populations in the field did increase in the manure treatment and were significantly greater than in the NH_4NO_3 treatment. Again, this may reflect differences in the quality of organic resources that were added to the manure treatment in the field and lab. The manure added to the field plots contained considerably more straw than that added to the microcosm. Field plots that received manure also received organic inputs as weed biomass (940 kg ha^{-2}) and had a standing stock of coarse organic matter (2800 kg ha^{-2}, data from Ketterings, 1992) that was removed from soil used in the microcosms. These organic materials probably provided substrates for fungal growth than were absent in the microcosms. Fungal populations were not evaluated in the current study, but straw residues are known to support substantial fungal growth (Eitminaviciute et al., 1976; Wessen & Berg, 1986; Broder & Wagner, 1988).

Population levels of plant parasitic nematodes in our experiments were low compared with those reported in other studies (Thomas, 1978; Baird & Bernard, 1984; Niblack & Bernard, 1985; Parmelee & Alston, 1986). This may be because the field site had received regular applications of cattle slurry prior to the establishment of our field experiment. There is evidence that organic amendments can suppress populations of plant parasitic nematodes (Mankau, 1980; Singh & Sitaramaiah, 1973; Mushkoor et al., 1977).

Because of the importance of biological interactions to soil fertility (Hendrix et al., 1986; Edwards & Stinner, 1988; Beare et al., 1992), it is important to understand the differences between detrital food webs that develop on different organic residues and under different soil conditions. Agroecosystems contain or are amended with many different forms of organic matter (coarse and fine particulate organic matter, crop and weed residues, cover crops, animal manures), which may be managed in many different ways (plowing, no-tillage, conservation tillage, herbicides, mulches, manures). Different residues support different microbial communities (Broder & Wagner, 1988), which, in turn, influence trophic communities and nutrient dynamics (Heal & Dighton, 1985; Parmelee et al., 1989; Wasilewska & Bienkowska, 1985; Beare et al., 1992).

In our experiments, the plant residue treatment supported a greater population and proportion of fungivores than did the animal manure treatment. Populations of bacterivores increased greatly and were not significantly different in these two organic treatments. Fungivores were relatively more abundant in the field than in the microcosm experiment. Differences that were observed in the response of nematode functional groups to the three different inputs used in these experiments, and under the different conditions of the field and microcosm experiments, were due mainly to differences in the quality and abundance of organic resources. Future studies will include analyses of microbial communities so that the underlying mechanism of these responses and their impact on nutrient dynamics in these systems is better understood.

SUMMARY

The trophic structure of free-living nematode communities responds to soil inputs and management practices and can be used as a measure of the biological response of soils to these manipulations. The purpose of this study was to determine the short-term responses of nematode communities to one inorganic and two organic nutrient inputs in field and laboratory microcosm studies. In the field experiment, the nutrient inputs were NH_4NO_3 fertilizer, a hairy vetch/rye cover crop, or straw-packed cow manure, which were added at a rate of 100 kg N ha^{-1} to corn agroecosystems in Wooster, OH. Nematodes were sampled monthly in the growing season. The microcosms contained 16 L of sieved soil from the field site. Amendments that were similar to those used in the field study were added to the microcosms at a rate of 150 kg N ha^{-1}. Nematode communities in the microcosms were sampled over 4 mo. The legume treatment led to the greatest increase in fungivorous nematodes in both field and microcosm studies. In both field and lab studies, populations of bacterivorous nematodes increased greatly in both the legume and manure treatments, but not in the NH_4NO_3 treatment. Fungivore populations did not increase in response to inputs of manure and NH_4NO_3 in the microcosm study, but in the field study, were significantly greater in the manure than in the NH_4NO_3 treatment. The differences in response of the nematode communities among different nutrient treatments and between the field and microcosm studies were attributed to differences in the quality and availability of organic inputs.

ACKNOWLEDGMENTS

The authors thank Dr. R.W. Parmelee for critical comments on the manuscript.

REFERENCES

Alejnikova, M.M., T.I. Artemjeva, T.M. Borisovic, F.G. Gatilova, S.M. Samosova, N.M. Utrobina, L.I. Sitova. 1975. Successions of microorganisms and invertebrata and their connections with biochemical processes during decomposition of manure in soil. Pedobiologia 15:81–87.

Baird, S.M., and E.C. Bernard. 1984. Nematode population and community dynamics in soybean-wheat cropping and tillage regimes. J. Nematol. 16:379–386.

Beare, M.H., R.W. Parmelee, P.F. Hendrix, W. Cheng, D.C. Coleman, and D.A. Crossley, Jr. 1992. Microbial and faunal interactions and effects on litter nitrogen and decomposition in agroecosystems. Ecol. Monogr. 62:569–591.

Bernard, E.C. 1992. Soil nematode biodiversity. Biol. Fertil. Soils 14:99–103.

Bongers, T. 1990. The maturity index: An ecological measure of environmental disturbance based on nematode species composition. Oecologia 83:14–19.

Bostrom, S., and B. Sohlenius. 1986. Short-term dynamics of nematode communities in arable soil: Influence of a perennial and an annual cropping system. Pedobiologia 29:345–357.

Broder, M.W., and G.H. Wagner. 1988. Microbial colonization and decomposition of corn, wheat, and soybean residue. Soil Sci. Soc. Am. J. 52:112–117.

Clarholm, M., B. Popovic, T. Rosswall, B. Soderstrom, B. Sohlenius, H. Staff, and A. Wiren. 1981. Biological aspects of nitrogen mineralization in humus from a pine forest podsol incubated under different moisture and temperature conditions. Oikos 37:137–145.

Coleman, D.C., R.V. Anderson, C.V. Cole, J.F. McClellan, L.W. Woods, J.A. Trofymow, and E.T. Elliott. 1984. Roles of protozoa and nematodes in nutrient cycling. p. 17–28. In R.L. Todd and J.E. Giddens (ed.) Microbial–plant interactions. ASA Spec. Publ. 47. ASA, CSSA, and SSSA, Madison, WI.

Edwards, C.A., and B.R. Stinner. 1988. Interaction between soil-inhabiting invertebrates and microorganisms in relation to plant growth and ecosystem processes. Agric. Ecosyst. Environ. 24:1–3.

Eitminaviciute, I., Z. Bagdanaviciene, B. Kadyte, L. Lazauskiene, and I. Sukackiene. 1976. Characteristic successions of microorganisms and soil invertebrates in the decomposition process of straw and lupine. Pedobiologia 16:106–115.

Freckman, D.W. 1988. Bacterivorous nematodes and organic-matter decomposition. Agric. Ecosyst. Environ. 24:195–217.

Freckman, D.W., and E.P. Caswell. 1985. The ecology of nematodes in agroecosystems. Annu. Rev. Phytopathol. 23:275–296.

Golebiowska, J., and L. Ryszowski. 1977. Energy and C fluxes in soil compartments of agroecosystems. Ecol. Bull. 25:274–283.

Heal, O.W., and J. Dighton. 1985. Resource quality and trophic structure. p. 333–354. In A.H. Fitter (ed.) Ecological interactions in soil: Plants, microbes and animals. Blackwell Scientific Publ., Oxford, England.

Hendrix, P.F., R.W. Parmelee, D.A. Crossley, Jr., D.C. Coleman, E.P. Odum, and P.M. Groffman. 1986. Detritus foodwebs in conventional and no-tillage agroecosystems. BioScience 36:374–380.

Huhta, V., and H. Setala. 1990. Laboratory design to simulate complexity of forest floor for studying the role of fauna in the soil processes. Biol. Fertil. Soils 10:155–162.

Ingham, R.E., J.A. Trofymow, E.R. Ingham, and D.C. Coleman. 1985. Interactions of bacteria, fungi, and their nematode grazers: Effects on nutrient cycling and plant growth. Ecol. Monogr. 5:119–140.

Ketterings, Q.M. 1992. Effects of earthworm activities on soil structural stability and soil carbon and nitrogen storage in organic-based and conventional agroecosystems. M.S. thesis. Wageningen Agricultural Univ., Wageningen, the Netherlands.

Mankau, R. 1980. Biological control of nematode pests by natural enemies. Annu. Rev. Phytopathol. 18:415–440.

McSorley, R., and E.E. Walter. 1991. Comparison of soil extraction methods for nematodes and microarthropods. p. 201–208. In D.A. Crossley, Jr. et al. (ed.) Modern techniques in soil ecology. Elsevier, New York.

Mushkoor, A.M., S.A. Siddiqui, and A.M. Khan. 1977. Mechanism of control of plant parasitic nematodes as a result of the application of organic amendments to the soil: III. Role of phenols and amino acids in host roots. Indian J. Nematol. 7:27–31.

Neher, D. 1992. Ecological sustainability in agricultural systems: Definition and measurement. p. 51–62. In R.K. Olson (ed.) Integrating sustainable agriculture, ecology, and environmental policy. Proc. Workshop USEPA, Arlington VA. July 1991. Food Products Press, New York.

Niblack, T.L., and E.C. Bernard. 1985. Nematode community structure in dogwood, maple and peach nurseries in Tennessee. J. Nematol. 17:126–131.

Parmelee, R.W., and D.G. Alston. 1986. Nematode trophic structure in conventional and no-tillage agroecosystems. J. Nematol. 18:403–407.

Parmelee, R.W., M.H. Beare, and J.M. Blair. 1989. Decomposition and nitrogen dynamics of surface weed residues in no-tillage agroecosystems under drought conditions: Influence of resource quality on the decomposer community. Soil Biol. Biochem. 21:97–103.

SAS Institute. 1985. SAS user's guide: Statistics. SAS Inst., Cary, NC.

Singh, R.S., and K. Sitaramaiah. 1973. Control of plant parasitic nematodes with organic amendments of soil. Exp. Stn. Res. Bull. 6. Univ. Agric. Tech., Pantnagar, India.

Sohlenius, B., and S. Bostrom. 1984. Colonization, population development and metabolic activity of nematodes in buried barley straw. Pedobiologia 27:67–78.

Sohlenius, B., S. Bostrom, and A. Sandor. 1987. Long-term dynamics of nematode communities in arable soil under cropping systems. J. Appl. Ecol. 24:131–144.

Sohlenius, B., and A. Sandor. 1989. Ploughing of a perennial grass ley; effects on the nematode fauna. Pedobiologia 33:199–210.

Stork, N.E., and P. Eggleton. 1992. Invertebrates as determinants and indicators of soil quality. Am. J. Altern. Agric. 7:38–47.

Sudhaus, W., K. Rehfeld, D. Schluter, and J. Schweiger. 1988. Interrelationships of nematodes, beetles and flies in the succession of cow pats during decomposition. Pedobiologia 31:305–322.

Thomas, S.H., 1978. Population densities of nematodes under seven tillage regimes. J. Nematol. 10:24–27.

Wasilewska, L. 1979. The structure and function of soil nematode communities in natural ecosystems and agrocenoses. Polish Ecol. Studies 5:97–145.

Wasilewksa, L., and P. Bienkowska. 1985. Experimental study on the occurrence and activity of soil nematodes in decomposition of plant material. Pedobiologia 28:41–57.

Wasilewska, L., E. Paplinska, and J. Zielinski. 1981. The role of nematodes in decomposition of plant material in a rye field. Pedobiologia 21:182–191.

Wessen, B., and B. Berg. 1986. Long-term decomposition of barley straw: Some chemical changes and ingrowth of fungal mycelium. Soil Biol. Biochem. 18:53–59.

Yeates, G.W. 1971. Feeding types and feeding groups in plant and soil nematodes. Pedobiologia 11:173–179.

Yeates, G.W., and D.C. Coleman. 1982. Nematodes in decomposition. p. 55–80. In D.W. Freckman (ed.) Nematodes in soil ecosystems. Univ. of Texas Press, Austin, TX.